T0221189

Building Information Modeling

Building Information Modeling

A Strategic Implementation Guide for Architects, Engineers, Constructors, and Real Estate Asset Managers

Dana K. Smith, FAIA
Michael Tardif, Assoc.
AIA, CSI, Hon. SDA

WILEY

John Wiley & Sons, Inc.

Library of Congress Cataloging-in-Publication Data:

Smith, Dana K.
  Building information modeling : a strategic implementation guide for architects, engineers,
constructors, and real estate asset managers / Dana K. Smith, Michael Tardif.
    p. cm.
  Includes bibliographical references and index.
  ISBN 978-0-470-25003-7 (cloth)
    1. Building information modeling. I. Tardif, Michael. II. Title.
TH437.S555 2009
692—dc22
                                                                    2008055951

*For my Mom and Dad, who instilled in me the will to persevere and not see any
task as impossible; for my wife, Patricia, who has put up with me
all these years, and for Michael, without whom this book would not
have become a reality.*

—Deke

*For all of our colleagues in the building industry who refuse to accept the status
quo and who share our passion to build a better world.*

—Michael

CONTENTS

Foreword   xi

Introduction   xv

CHAPTER 1   Building Industry Challenges and Opportunities   1

             Global Trends in Supply and Demand   2
             Benchmarking Construction Productivity   3
             Construction Productivity Metrics   6
             Benchmarking Building Performance   8
             Converting Inefficiency and Waste into Profit   10
             Benchmarking Waste   11
             Identifying Business Opportunities   12
             Emerging Business Strategies   15
             Choosing the Right Tools, Deploying the Right
             Tool Suites   16
             The BIM Value Proposition   19
             Process Engineering   20
             Thinking Like an Owner   22
             Building Performance Metrics   23
             New Metrics for Real Property Valuation   23

CHAPTER 2   BIM Implementation Strategies   27

             Leaving the CAD Era Behind   28
             A Systems Approach to BIM Implementation   29
             Avoiding Ideological Pitfalls   30
             Aligning a BIM Implementation Strategy with
             Technology Trends   32
             Assessing Fundamental Risks   33
             Fostering a Culture of Information Stewardship   33
             Managing Culture Change   35
             Using Technology to Build Trust and Mitigate Risk 36
             Maintaining Data Exchange Capabilities   37
             Assessing Team Capabilities   41
             Managing Expectations   42
             Measuring Progress toward Strategic Goals   44
             Toward a New Business Paradigm   54

CHAPTER 3   Business Process Reform   57

Managing Innovation Risk   58
The Imperative of Change   59
Innovation Management Strategies   60
The "I" in BIM   62
Business Reform Strategies   63
Industry-wide Reform Efforts   64
Industry Standards and Innovation   66
The Industry Standards Landscape   68
Aligning Business Strategies with Industry
Standards   70
Integrating Information Gathering into the
Business Process   72
Leadership and Vision   75
Engaging Business Partners   77
Business Process Modeling   78
Business Process Modeling Case Study   82
Managing Change   83
Populating the Building Information Model   87

CHAPTER 4   BIM-Based Enterprise Workflow   89

BIM Implementation Fundamentals   90
Sidebar: Integrating Data Collection with
Business Processes   94
Business Operations and BIM   98
Marketing/Business Development   99
Human Resources   101
Finance   102
Information Technology   104
Operations   105
Workflow Visualization   108

CHAPTER 5   The Building Life Cycle   111

Life Cycle Views of Building Information   111
The Feasibility, Planning, and Development View   115
The Design and Construction View   116
The Operations and Maintenance View   118
The Ownership and Asset Management View   123

CHAPTER 6   Building Information Exchange Challenges   129

Sidebar: Data, Information, Knowledge, and Wisdom   130
Information Management   131
Sidebar: Case Study: Information Management   134

Information Provenance   135
Information Maturity   137
Sidebar: The Wayback Machine: Archiving the Web   138
Information Content Decay   140
Information Electronic Degradation   140
Information Integrity and Continuity   141
Information Transparency, Accessibility, and Security   143
Information Flow   144
The Life Cycle of Information   146
Stakeholder Views   146
Interoperability   146

CHAPTER 7   Building Information Exchange Requirements   153

The Big Picture   154
Information Delivery Manuals   154
Defining "Best Case" Business Processes   159
agcXML: Organizing Transactional Information   161
SMARTcodes: Automating the Regulatory Process   162
The Construction Operations Building Information
Exchange   163
Specifiers Property Information Exchange (SPIE)   166
Coordination View Information Exchange (CVIE)   168

CHAPTER 8   The Way Forward   171

Workflow: From Sequential to Parallel Processing   173
Business and Contractual Relationships   176
Evolving Roles and Responsibilities   177

Bibliography   179
Index   183

Foreword

"It was the best of times, it was the worst of times. . ." Charles Dickens could not have said it better. Although he was speaking of the French Revolution, the same applies to our time in the first decade of the twenty-first century. Economic upheaval has spread around the globe, making everyone concerned about the future. But that is not the worst of times I am referring to. I refer to the complexity of the building industry, and the difficulty with which projects are brought to fruition due to the industry's fractured, adversarial nature.

However, amidst this seeming battleground of competing interests lies tremendous opportunity, thus making it the best of times. In this book, the authors, Dana K. Smith and Michael Tardif, identify, capture, and define this opportunity with tremendous clarity and completeness through their understanding of the available emerging technology and how best to use it.

Building Information Modeling, or BIM, is not a wave of the future—it is happening right now. BIM is utilizing Computer Aided Design (CAD) technology in a way that ultimately will tie all the components of a building together as objects imbedded with information. A simple computer graphic turns into an object pregnant with information that tracks its manufacture, cost, delivery, installation methods and labor costs, and maintenance all the way through to its replacement value. More importantly, BIM makes possible the erection of the building in computer model form before putting a shovel in the ground. This goes a long way to identifying and solving the typical errors that occur during construction because drawings were not coordinated properly during their creation. Fixing a problem on the computer model is a fraction of what it would cost to fix the mistake in the field.

In *Building Information Modeling: A Strategic Implementation Guide for Architects, Engineers, Constructors, and Real Estate Asset Managers*, the authors address the wonderful potential of BIM without ignoring its flaws, giving the reader a better understanding of how to introduce the technology to their practice in a practical way while still holding out the very exciting promise of what is possible in the near future. This pragmatic approach to the present and perceptive look into the possibilities of the future give the book a fascinating tension that drives the reader onward, eager to learn more.

On the practical side, BIM technology has begun to do for the building industry what has been needed for a long, long time—enabling the measurement of building performance metrics, especially the incorporation of information technology, and how it has changed the way we design and build our structures over the last 20–30 years. Right from the start, this could have a tremendous impact for it is estimated that the $1.288 trillion building industry marketplace wastes over 57 percent of that, according to a 2004 study (James E. Diekmann et al., "Application of Lean Manufacturing Principles to Construction," Austin, Texas: Construction Industry Institute, 2004). That means over $600 billion is lost each year. Capturing even a small percentage of this would be significant, and the authors believe that BIM will be successful in doing that and more.

BIM also holds many promises yet to be realized. Although the authors do not spend a great deal of time on these issues, there is still a powerful sense that BIM, utilized to its fullest extent, can migrate "upstream" from the building design and construction process to the modeling of the management and operations phases that the building contains. This would totally model the entire enterprise and allow the architecture profession to assert a greater leadership role throughout the life cycle of buildings. Architects would be much more helpful in identifying how operational processes should be laid out, where the building should be located in relation to markets and labor resources, and how it would maximize the productivity of the client once it is built. By integrating both the operational and structural aspects of the client's enterprise through what could be called Enterprise Design, BIM enables the whole to be better than the sum of its parts. Also, the architect that uses BIM to integrate work flow with work product will realize a better result both in performance and pay, making the architect's role more profitable. But this is the next book I challenge the authors to write!

Right now, BIM is transforming the way the built environment is being created and sustained. This book offers one of the most comprehensive discussions of how this is happening and why this is so important. Good design will always trump project management, but with BIM, good design will be made even better and more cost effective. In these times of terrible financial uncertainty those are comforting words, indeed. Anything that can bring order, understanding, and efficiency into an industry fraught with conflict, complexity, and waste will certainly turn the worst of times into the best of times. All you need to do is read on to find out this can be done by implementing Building Information Modeling in your office.

Ambassador Richard Swett, FAIA

Ambassador Swett is a licensed architect who served as U.S. Ambassador to Denmark and as Representative to the U.S. House of Representatives from the state of New Hampshire. He is the former Vice President and Managing Principal of the Washington, D.C. office of LEO A DALY Architects and Engineers and is currently president of the New Hampshire-based consulting firm Swett Associates. He is also the author of *Leadership by Design: Creating an Architecture of Trust*.

Introduction

In the December 16, 2002, issue of the LaiserinLetter™, in what may have been the first widely published article on the topic, industry analyst Jerry Laiserin weighed in on the debate over a new term or acronym to describe the newly emerging design technology then poised to replace computer-aided design (CAD).[1] Laiserin opined that the lack of a clear, meaningful term for this new technology was "a deadly serious issue that [if left unresolved] can stymie meaningful discussion." Citing a recent meeting of building industry strategists in which more than half of the scheduled meeting time had been devoted to crafting a term that all attendees could agree upon, Laiserin made a cogent argument for the term "building information modeling," or BIM, as the best term to describe "the next generation of design software." Many have since attributed authorship of the phrase to Laiserin, a misconception he definitively corrects in his introduction to the *BIM Handbook* by providing the most complete written account to date of the term's evolution. There can be little doubt, though, that Laiserin's 2002 article marks the point at which the term "BIM" first came into popular use.

The new acronym addresses the problem that Laiserin articulated by differentiating the digital design tools available to the building industry today from the CAD tools first developed a generation earlier. But the debate over the precise meaning of the term continues. In countless meetings and industry forums, far too much time is still devoted to defining exactly what the terms "building information model" and "building information modeling" really mean.

The continuing confusion, as we see it, can be attributed to the multiple meanings of the word "model." In his 2002 article, Laiserin made note of these multiple, precise, and useful meanings, among them "the mathematical or digital description of objects or [complex] systems, [such as] econometric models and weather models, as well as physical models of 3-D objects." He also noted that the verb form *to model*, which building design professionals commonly construe as the act of constructing a physical, scaled model, "also implies a process of . . . building performance simulation (essentially, modeling future behavior)."

Regrettably, this most useful, mathematical sense of the term—the *modeling* of processes or systems—has failed to lodge itself in the collective consciousness of design professionals, who reflexively associate the word *model* (and by extension, BIM) with physical geometry and not processes. Architects and engineers tend to think of BIM primarily as a tool for creating scaled, three-dimensional, virtual representations of actual buildings. This is reinforced by the design of most BIM authoring applications, which offer a user interface optimized for the creation of three-dimensional building geometry at the expense of other data input methods. Only recently—as constructors have begun using BIM to assess the constructability and construction sequencing of buildings prior to construction—has the building industry begun to appreciate the full potential of BIM for compiling vast amounts of building information other than geometry and for radically altering the way in which buildings are designed, built, operated, and maintained throughout their useful lives.

To be fair, most design professionals immediately recognize the qualitative difference between the "intelligent" geometry of a building information model and the much simpler geometry generated by a CAD application or 3-D visualization tool. CAD applications use trigonometric algorithms to define vectors (three-dimensional lines) and simple geometric shapes such as cylinders, arcs, and cones. The end product is a wire-frame digital model of a building. Visualization tools may go a step further, combining vectors, raster imaging, and solid-surface modeling to generate realistic 2-D or 3-D digital representations of buildings. But in either case, the considerable mathematical legerdemain needed to depict the actual components of a building has little if anything to do with the body of knowledge needed to design and construct the building itself. Mastering these applications often meant having to know as much about the underlying algorithms as one did about architecture, engineering, or construction. The need to acquire that computational knowledge was—and is—a costly burden that has added little value to the products and services of design and construction professionals. Instead, it has imposed a very real though often hidden overhead cost on the entire industry.

A BIM application, by contrast, more readily "understands" that the objects created by users represent real-world components of actual buildings, such as doors, windows, walls, and roofs. The underlying algorithms that define these objects are no less complex than those developed for CAD applications, but with BIM the software design and computation effort is directed toward creating a user experience that is much closer to reality for design and construction professionals, if not yet for the entire building industry. With BIM, a greater percentage of intellectual bandwidth can be devoted to the object of professional attention—the building—than is possible with CAD. BIM objects can be imbued with the characteristics of their real-world counterparts, so that, for example, a window "knows" that it can exist only in a wall, and a wall "understands" that one of its essential attributes is thickness. This embodied knowledge is not so much intelligence as it is common sense about how these objects are expected to behave in the real world. But such "intelligent objects" are what distinguishes the geometry created by a building information modeling application from a mere 3-D model. And yet, even the revolutionary concept of intelligent objects does not move us beyond a very limited conception of BIM as intelligent geometry.

The geometry of a building represents only a small percentage of the total body of useful information about that building. A genuinely comprehensive building information model would encompass not only geometry but all of the information about a building that is created throughout its useful life. The core technology would have more in common with a relational database than a CAD application. The building information would be accessible to many different types of users—building owners, operators, constructors, facility managers, portfolio managers, and even emergency responders—through user interfaces that are accessible and familiar to each.

The primary focus on geometry in first-generation BIM applications is perhaps unavoidable. Relational database software architecture—as complex as it can be—is built upon the presumption that the data is alphanumeric; that it consists of words and numbers. It is relatively easy, from a software design point of view, to maintain a bright line in a relational database between the data itself and the software that is used to act upon it. It is nearly impossible, however, to separate geometric data from the complex algorithms that are used to create or display it. This makes it extremely challenging to compile both geometrical and alphanumeric building information into a single relational database, or to exchange building information reliably among dissimilar software applications.

The AEC software industry, however, often has done more to confuse than to clarify the issue. Not wanting to be perceived as purveyors of outdated technology, building design software developers have been quick to embrace BIM as a moniker regardless of the actual capabilities or underlying technology of their applications, which, not surprisingly, are most adept at producing geometry and visualization, whether intelligent or not. It can be exceedingly difficult for an end user, when presented with comparable images of sophisticated three-dimensional models, to understand fully just how sophisticated and intelligent the underlying data really is. This has created an environment in which companies with truly innovative technology, in an effort to distinguish their "true" BIM applications from those that have been merely branded as "BIM software," introduced new terms into the lexicon, such as "4-D," which is generally understood to mean "3-D + time," or "macro BIM," whose meaning we will not even attempt to explain. Not to be outdone, others have advanced the notion of "5-D" and "6-D" modeling, where the ever-increasing numbers refer to other "dimensions" (translation: attributes) of buildings, such as their cost, the sequence of construction over time, or even the total embodied energy needed to build them. Given the difficulty we have had thus far broadening our understanding of the word "model," introducing this new meaning of the word "dimension" is not likely to help matters very much.

Changing the way we think about BIM means changing the way we think about building information and the *modeling* of building information, which actually has very little to do with technology. We have been generating building information in a more or less organized fashion for over a hundred years—in construction drawings, specifications, scale models, building component schedules, product data sheets, schedules of values, building permits, requests for information, change orders, bills of material, data logs, purchase orders, invoices, applications for payment, field reports, punch lists, certificates of occupancy, photographs, as-built drawings, commissioning procedures, operation and maintenance manuals, warranties, material disposal certificates, and dozens of other documents. What we have failed to do is compile these unwieldy paper records into a single information repository. We are now at risk of repeating the same mistake with digital building information.

Visualizing a building information model as a compendium of these familiar forms of data is easier and more useful than trying to visualize "nD" building information. It allows us to grasp the full scope of information that a digital building information model might contain without having to understand anything about technology. It puts the focus back on *information.* It reminds us that the bulk of building information is alphanumeric data and not geometry. It illustrates that the primary challenge facing the industry is not figuring out how to deploy new technology but how to do a better job of organizing and exchanging the information we create every day.

Many early proponents of BIM—ourselves included—attempted to highlight the central place of information in *building information modeling* in a literal way by articulating a vision of a "single building model" that could be "developed and sustained throughout the life cycle of a building facility." This was commonly understood to mean that all of the available information about a building would be contained in a single electronic data repository stored on a secure Web server accessible to any party who might need to query, add to, or remove information from the model at any time in the life cycle of the building. Anyone who has attended an industry conference in the last few years has seen the illustration of two workflow models—existing and proposed—invariably used to illustrate the single building model concept. The "existing workflow" diagram shows all of the parties involved with a building throughout its life cycle as small circles arranged about an open circle, with lines extending from each party to every other party in a dense (and what is intended to be a self-evidently hopeless) web of communication and information exchange. The "proposed workflow" diagram shows the same arrangement of the parties, but in the center is a large circle marked "BIM," or "single building model," with a single line of communication radiating outward from the oracular BIM to each orbiting party like the spokes of a wheel. Implicit in this concept is that the model is comprehensive, accessible, and accurate at all times.

Anyone with a foot in the real world can instantly tick off a long list of challenges to the business model depicted in the second diagram, beginning with the stewardship of the model: Who owns it? Who maintains it? Who is responsible for it? Who pays for it? Attempts to answer those questions raise still more questions about the accountability, responsibility, and liability of the parties involved; the reliability, security, and integrity of the data; and the data storage and data access infrastructure needed to support it. It soon becomes clear that the proposed single building model raises a host of business problems without providing any clear, demonstrable benefits.

The way out of the dilemma, again, is to broaden our understanding of BIM to focus not on the data but on the business processes used to create it—on the *modeling*, not the *model*. Creating a single building model is not the desired outcome of BIM; instead, the goal is compiling comprehensive, reliable, accessible, and easily exchangeable building information for anyone who needs it throughout the life cycle of a building. The differences may be subtle but are nonetheless important. One approach places the highest value on the data itself; the other places the highest value on the purposes for which the data can be used.

The concept of the single building model illustrates the danger that ideology can pose to real progress. The simplicity of the concept is alluring, but it doesn't solve any real problems or align with any real business processes. The current messy web of bilateral communication and information exchange in the building industry is necessary—people need to be able to talk to and exchange particular pieces of information with other particular people at particular times for particular purposes. This simple fact will not change, nor should it.

The best technologies do not try to replace vital lines of communication and workflow. Instead, they streamline workflow by minimizing or eliminating routine, non-value-added tasks and maximizing high-value-added tasks. The seasoned judgment of building industry professionals—which many feared would be supplanted by technology—is instead being enhanced and leveraged by technology.

The concept of the single building model also runs counter to another simple fact: no one ever needs all of the information about a building at any one time. The value of different pieces of information—and with it, the commitment of various parties to collecting and preserving it—varies widely throughout the life cycle of a building. Compiling and maintaining building information in a single repository for the life cycle of a building is useful to individual players only to the extent that it supports the many individual business processes throughout the life cycle. BIM technology and workflow must be aligned in such a way as to foster routine data collection and preservation, not add it as an additional burden. The information life cycle goals of BIM proponents will be realized only if these obvious facts are taken into account.

In a more recent version of the "two workflows" illustration, as shown in Figure 0.1 the object at the center of the proposed workflow diagram—formerly labeled "single building model," is now labeled "industry exchange standard." This subtle change illustrates a substantially different and far more mature technology concept. The notion here is not that all data is to be contained in a single building model but that all of the parties in the building life cycle will exchange information through a single information exchange standard. The diagram depicts an environment of true interoperability in which each party uses whichever software tool is best suited to a particular task, but every party shares information reliably with every other party through a common information exchange protocol. In a study completed by the Open Application Group (OAGi) for the Open Standards Consortium for Real Estate (OSCRE), the cost of developing bilateral information exchange mechanisms to support "full interoperability" among twenty different software applications was found to be

FIGURE 0.1
BIM: An Interop-
erability View.

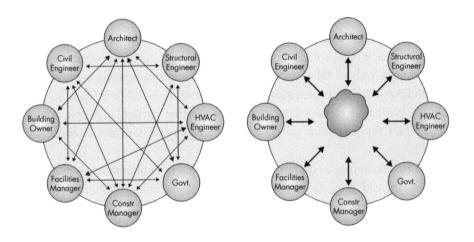

twenty times higher than the cost of developing a single information exchange mechanism for all twenty applications.[2] As additional applications are added to the mix, the cost of developing bilateral exchange mechanisms with all other applications rises exponentially, while the cost of incorporating support for the common information exchange mechanism rises only arithmetically.

The "single building model" version of the illustration reminds us that we should not allow seemingly elegant aspirational goals of BIM to get in the way of the real business goals, while the "interoperability" version underscores just how important collaboration and cooperation across the building industry really is.

Greater productivity and efficiency across the entire life cycle of any building is the foremost reason for deploying new technology. We need to develop faster, less expensive, better, and more efficient ways to design, build, manage, operate, use, maintain, repair, and demolish or reuse buildings. If technology concepts such as the single building model turn out to be the best way to achieve those goals, so much the better. If not, we should be prepared to jettison them for more workable solutions or refine our thinking to evolve new concepts.

Business objectives will determine how BIM technology develops. We have no shortage of terms to describe the powerful business drivers: integrated project delivery; scenario planning; rapid, iterative design; virtual design and construction; lean construction; value engineering; sustainability; real property asset management; preventative maintenance; energy conservation; environmental stewardship; life cycle costing. BIM is an enabling technology for achieving these objectives. It is a means to an end, not an end itself.

So we want to be clear, from the outset, what the terms *model* and *modeling* mean to us and how we will use them in this book. *Any* compilation of building information, in any form, is a building information model. *Any* simulation of *any* real activity related to a building is an act of *building information modeling*. As Kimon Onuma, FAIA, an architect and software developer in Pasedena, California, has pointed out, a spreadsheet of spatial data is a *building information model,* and if you use the alphanumeric data contained in the spreadsheet to simulate an actual business process in the life cycle of a building, such as asset management, then you are engaged in *building information modeling*. There is no minimum or maximum number of variables or attributes (dimensions) that can be considered. The "nD" phenomenon is an integral part of the original meaning of BIM, not something beyond BIM.

Building information modeling is nothing more—and nothing less—than a *systems approach* to the design, construction, ownership, management, operation, maintenance, use, and demolition or reuse of buildings. A *building information model* is any compilation of reliable data—in single or multiple electronic data formats, however complete or incomplete—that supports a systems approach in any stage in the lifecycle of a building. The implications of these simple (some may say simplistic) definitions are enormous. A systems approach requires that everyone in the industry, *though acting independently,* must begin thinking of themselves, their products, and their services as part of a *system.* We've observed that when this change of mindset occurs, interesting things start to happen. Companies begin to collaborate. Data exchange challenges are overcome. Adversarial or punitive contract clauses—which create artificial project risks—are replaced by mutual indemnification and performance incentive clauses—which reduce real project risks. Non-value-added tasks are automated. Defensive, irrelevant documentation is eliminated. Design quality is enhanced. Waste is reduced. Productivity improves. Costs go down. Profits go up. And everyone recognizes and exploits opportunities to improve the products they make, enhance the services they deliver, or expand their market share. In the process, some players invent entirely new ways of doing things. That, we believe, is the true potential of BIM. Ironically, we didn't really need a computer to accomplish these business goals, but the technology is finally providing the needed impetus for business reform that should have occurred long ago.

Dana K. Smith, FAIA
Herndon, Virginia
Michael Tardif, Assoc. AIA, CSI, Hon. SDA
Bethesda, Maryland

ENDNOTES

[1]Jerry Laiserin, "Comparing Pommes and Naranjas," *LaiserinLetter,* no. 15 (2002), www.laiserin.com.

[2]Andy Fuhrman, "OSCRE Update: The OSCRE Standard for Property E-Commerce,: in *American Institute of Architects Building Connections* 3 (Washington, DC: The American Institute of Architects, 2006).

CHAPTER 1

Building Industry Challenges and Opportunities

Until one is committed, there is hesitancy, the chance to draw
back. . . .
Whatever you can do, or dream you can do, begin it.
Boldness has genius, power, and magic in it. Begin it now.

—Goethe

The building industry is facing a looming worldwide crisis, a spectacular convergence of gross inefficiency and inordinate consumption of energy and raw materials. While the spectre of global warming has become a catalyst for renewed interest in conserving energy and raw materials throughout the life cycle of buildings, the environmental challenge only adds greater urgency to a far more elemental problem: the utter failure of the building industry to keep pace with the technological advancements and productivity gains of nearly every other industry in the last fifty years. Even farming, the most ancient productive activity in human civilization, has managed to achieve productivity gains in the last hundred years that are unimaginable in the building industry.

Technological advancement is measured largely by increases in efficiency, whether in means and methods of production or the consumption of raw materials. Increased efficiency and productivity lower costs, increase profits, and help raise the standard of living by making goods and services affordable to greater numbers of people. By that measure, the worldwide building industry has accomplished very little in the way of technological advancement.

GLOBAL TRENDS IN SUPPLY AND DEMAND

An estimated 40 percent of global raw materials are consumed by building construction.[1] In the United States, when all other manmade, immovable structures are included—things such as bridges, roads, dams, and ports—the raw materials consumed by construction *exceeds 75 percent* of the total.[2] Or to put it another way, construction in the United States consumes three times more raw material than *all other economic and industrial activity combined.*

In 1900, each living American consumed two metric tons of raw materials per year. By 1995, the annual per capita figure had increased five-fold to 10 metric tons per person.[3] Global warming aside, simple economics dictate that we cannot sustain either our current standard of living or a growing U.S. population unless we reverse our steadily increasing rate of raw material consumption (see Figure 1.1). With construction accounting for three-quarters of the total, we cannot reverse the overall consumption trend unless we learn how to build more with less.

Both worldwide per capita consumption of raw materials and energy and the world's population are growing. These compounding trends will cause the worldwide demand for materials and energy to grow exponentially. Global population is expected to reach nine billion by 2050.[4] According to the United Nations, the world can sustain a population of only 1.8 billion at a high-income

FIGURE 1.1
Raw Materials
Consumption Chart.
(Source: U.S. Geological Survey.)

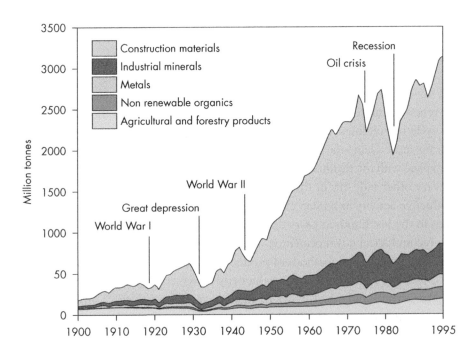

consumption level, a number we exceeded in 1965, and a population of just under 6 billion at a middle-income consumption level, which we passed in the mid-1990s.[5] If the 9 billion people living in 2050 are to have any hope of enjoying a reasonably decent standard of living, we will need to find ways to house, clothe, and feed ourselves—and support all of the economic activity we generate—far more efficiently than we do now.

Concern about worldwide standards of living is not a matter of altruism. Whether we like it or not, our economy is global. It is not possible to isolate ourselves from the global supply chain or the global supply and demand equation. The growing global demand for raw materials and energy resources will affect every consumer of construction materials in the United States. Demand for steel, lumber, and gypsum in China will affect the price Americans pay at home. The fundamental question is not whether we will run out of the raw materials we need; it is whether we will be able to afford them and continue to sustain a growing and profitable building industry.

So what's the good news? Simply this: in terms of efficiency and productivity, the building industry has nowhere to go but up. And given the inexorable growth in global population, a lean and efficient construction industry can count on a virtually limitless worldwide demand for its products and services for the foreseeable future. Successfully meeting that demand, however, will depend on whether the industry as a whole can deliver buildings of higher quality at a lower cost; buildings that are constructed with less energy and fewer raw materials, and generate less waste for their construction; buildings that are more durable and consume less energy while occupied; and buildings that are easier to recycle or adapt to new uses when they can no longer serve their original purpose. A failure to do so will lead to supply shortages or cost increases that global markets will be unable to bear, which will lead to economic contraction and widespread business failure.

BENCHMARKING CONSTRUCTION PRODUCTIVITY

Comparing the construction industry to agriculture over the last 50 years dramatically illustrates the point. From 1948 to 1999, the U.S. population grew from 147 million to just under 263 million, an increase of 79 percent.[6] Had the land needed for cultivation grown at the same rate as our consumption of raw materials, the amount of land we would need to feed ourselves would have had to increase nine-fold, from 1.6 billion acres in 1949 to 12.4 billion acres in 1999.[7] There's only one problem with that: the entire land area of the United States is only 2.3 billion acres. Had our rate of land consumption for growing food matched our rate of raw material consumption, we would have run out of land long ago. Instead, the amount of land under cultivation in the United States actually declined slightly from 1948 to 2004.[8] With a population that is now over 300 million, we are able to feed nearly twice as many people using the same amount of land as we did in 1948, and this does not even take into account that the United States is a net food exporter, so the gain in productivity is still greater.

Is the comparison fair? The per capita drop in land under cultivation would not have been possible without improvements in mechanized equipment and intensified use of pesticides, fertilizers, and other "inputs." If the total cost of all these inputs, including land and labor, had risen dramatically over the same period, then the per capita reduction in land use would not be an indicator of increased agricultural productivity. But according to the U.S. Department of Agriculture (USDA), while "the use of some inputs such as fertilizer and machinery has increased [between 1948 and 2004], these increases were more than offset by reductions in cropland and especially the amount of labor employed in agriculture."[9] In fact, the "total inputs" for agriculture (land, labor, capital, equipment, fertilizer, feed, and seed) have remained flat since 1948, while total output has increased 270 percent.[10] If the construction industry had achieved a comparable increase in productivity, a building constructed in 1948 at a cost of $1 million could have been built in 2004 (in constant 1948 dollars, before adjusting for inflation) for just $370,000.

Comparing the building industry to other industries—and looking for lessons from those industries that might be applied to our own—has become a popular pastime of late. We look at design automation and design-to-fabrication technologies in the automobile and aircraft industries and ask ourselves why we can't do the same thing. But there are two significant differences between construction and these other industries that make any comparison imperfect and any "lessons learned" difficult to apply. First, Boeing and Airbus are the only two commercial customers in the aircraft industry. They can dictate what they want from their suppliers, including exactly how goods and services should be delivered. They can effectively control their entire supply chain without the capital required to own it. The auto industry, while not as concentrated, is still dominated by a few global players who can define the "rules of engagement" with their suppliers. Second, both industries benefit from being able to assemble their products under factory conditions, giving them a great deal of control over the quality of the manufacturing process, the technology and capital that can be applied to it, and the supply and skill of labor.

The building industry, by comparison, is highly fragmented—there are millions of customers, end users, service providers, and product manufacturers.

No single entity commands sufficient market share to demand greater efficiency and productivity throughout the supply chain. Also, markets for many building industry goods and services tend to be local or regional, leaving consumers—building owners and facility managers—with a very limited range of product/ service options. Though up to one-third of building components are manufactured under factory conditions, most construction remains a field activity. The supply and quality of labor is more difficult to control. The weather affects working conditions and schedules. And despite an increase in the number, type, and variety of mechanized tools in the last fifty years, construction remains largely a craft process. A stone mason who worked on the Parthenon over 2,500 years ago could walk onto a job site today and, with little difficulty, recognize his tools and get to work. In this environment, the most that individual business enterprises can do is optimize their own operations, often to the detriment of overall efficiency and productivity.

Fair enough. But does the oft-repeated lament of industry fragmentation fully explain why the construction industry has failed to evolve technologically? The weakness of the argument becomes apparent when construction is compared to agriculture in greater depth. Though construction differs from the airline and automotive industries in important respects, the construction and agricultural industries of 1948 had many characteristics in common: a high degree of industry and market fragmentation; a diverse, poorly educated, and irregular supply of labor; limited access to sources of capital; little technology or automation; and very similar "field" working conditions. None of these characteristics impeded a rapid increase in productivity in agriculture. So how did agriculture manage to leap so far ahead of construction?

A lack of reliable statistical productivity data about the construction industry is a factor. Though the U.S. Department of Labor's Bureau of Labor Statistics (BLS) compiles productivity and unit labor cost data for most industries, "Productivity and Costs data are not published for any industries in construction."[11] We have nothing like the measure of total inputs and total outputs that the USDA uses to measure agricultural productivity.

The lack of such broad-based statistical data puts the construction industry at a significant disadvantage. If something as fundamental as productivity cannot be measured, then it is impossible to assess the effect on productivity—good or bad—that results from changes or improvements in technology, skill, business practices, or production methods. Nor is it possible to measure reliably whether productivity is going up or down over time.

When the impact of innovation cannot be measured, innovation is less likely to occur, and the risk of implementing change is much higher. It is often said that the building industry is conservative and resistant to change. One might argue, however, that the "resistance" actually reflects sound business judgment. The slow pace of innovation in the building industry is more likely the result of the lack of reliable business information. Among our highest priorities, then, should be to demand that the federal government compile the same detailed economic data that it compiles for every other industry in the economy and that all other industries take for granted.

CONSTRUCTION PRODUCTIVITY METRICS

Paul Teicholz, professor emeritus in the Department of Civil and Environmental Engineering and founding director of the Center for Integrated Facility Engineering at Stanford University, has advanced the idea of a "construction labor productivity index" that compares BLS data for fieldwork hours to U.S. Department of Commerce data for contract dollars of new construction (see Figure 1.2).[12] Teicholz does not measure productivity directly; he infers a measurement of productivity by comparing two unrelated macroeconomic data points compiled by two different U.S. Cabinet departments. But in the absence of more precise data, it may be the best available barometer of construction industry productivity.

FIGURE 1.2   Teicholz Construction Productivity Index Graph. Indexes of labor productivity for construction and non-farm industries, 1964–2004. Adapted from research by Paul Teicholz at the Center for Integrated Facility Engineering (CIFE) at Stanford University.

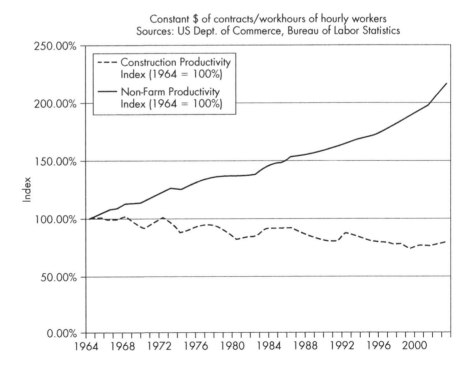

Constant $ of contracts/workhours of hourly workers
Sources: US Dept. of Commerce, Bureau of Labor Statistics

According to Teicholz, measuring constant contract dollars of new construction work per work hour reveals that productivity in the construction industry has *declined* by an average compound rate of 0.59 percent annually between 1964 and 2003, while labor productivity in all non-farm industries *increased* by 1.77 percent per year over the same period. Teicholz's index shows a cumulative drop in construction productivity of approximately 20 percent over 40 years.

Others in the construction industry have publicly challenged Teicholz's findings, but none of the dissenting opinions is based on the type of broad-based statistical economic data that makes similar measurements of productivity in other industries so reliable. Preston Haskell, chairman of the Haskell Company of Jacksonville, Florida, analyzed data of building construction projects completed by his company from 1966 to 2003 and concluded that real construction costs per square foot for four building types (warehouse, retail, office, and multifamily residential) had dropped by 12.3 percent over that period, while productivity had increased a total of 33 percent, or 0.78 percent per year.[13] Haskell's methodology is rigorous and well documented but is based on the data of a single company, which is no substitute for data gathered across an entire economic sector.

The core dilemma facing the building industry today is that the best data available—Teicholz's and Haskell's—varies so widely (more than 100 percent, from an estimated annual decline in productivity of 0.59 percent to an estimated increase of 0.78 percent) as to be of little or no value. We are left with inconclusive competing arguments rather than actionable statistical data.

Even if we could extrapolate Haskell's results to the industry as a whole, the annual increase in productivity that he claims his company has achieved is still considerably less than half that of all non-farm industries. Haskell attributes the difference—as does everyone else—to the fragmentation of the construction industry, where "research is nearly nonexistent because architects and engineers have neither the resources nor the incentive to fund research, and constructors have little ability to influence innovation in architectural, engineering, or product design."[14]

The key insight of Haskell's paper is not that the construction industry is fragmented, but that research plays a significant role in driving productivity. The agricultural and construction industries of 1948 were equally fragmented. Farmers had no greater financial resources than builders did and no greater incentives to fund research individually. The exponential and differential growth in agricultural productivity can be attributed almost entirely to the vast amounts of state and federal funding for agricultural research over the past sixty years, which has supported and continues to support a nationwide network of agricultural research stations managed by schools of agriculture at land-grant colleges and universities. Had the government funded research into construction at a comparable level over the same period, there can be little question that the industry could have achieved similar gains in productivity.

The growth in agricultural productivity is a worldwide phenomenon, at least in the developed world. This can be attributed, in part, to comparable investments in research by many national governments but also to the fact that the knowledge gained from government-supported research in agriculture is in the public domain and is widely published in academic and agricultural industry journals, which makes the knowledge available to anyone who wishes to apply it. As a result, innovation spreads rapidly.

The building industry is poised to move ahead with or without government support. However, the lack of research funding from private or public sources is acute, and if left unchanged will continue to hamper the pace of innovation. Industry organizations committed to change in the industry would do well to include advocacy of increased government funding for building performance research among their strategic planning goals.

In the absence of government funding, the importance of a global culture of innovation that includes knowledge sharing, open standards development processes, and full interoperability of digital building industry data cannot be overestimated. With such limited public investment in building industry research, private investment must be leveraged to achieve the greatest possible aggregate returns. Whenever a corporate enterprise, a software company, or an industry organization asserts proprietary control—as opposed to stewardship—of intellectual property in the realm of information exchange, the entire industry suffers, including the entity that originally seeks to protect its turf. The critical role of the culture of innovation in the building industry will be examined in greater depth in Chapter 3.

BENCHMARKING BUILDING PERFORMANCE

While productivity metrics are useful for assessing industry performance during the construction phase of buildings, energy efficiency is one of many metrics, and among the most easily quantifiable, for assessing the performance of buildings while they are in use. Energy efficiency statistics are as sobering as the productivity statistics and, according to at least one government agency, are not expected to improve.

In 2005, nonindustrial buildings accounted for 39.6 percent of all energy consumed in the U.S. and 71.8 percent of total U.S. electricity production.[15] (Industrial buildings are excluded from these figures due to the fact that the energy consumed by industrial buildings for building operations is indistinguishable from the energy consumed for the industrial operations that occur within them.) In its *Annual Energy Outlook 2007*, the U.S. Department of Energy (DOE) forecast that total U.S. energy consumption will increase 31 percent by 2030, from 100.2 to 131.2 quadrillion BTUs per year. Despite the many promising efforts underway to improve the energy efficiency of buildings, DOE forecasts that buildings will actually consume a marginally *greater* share of that higher total: 40.5 percent of total energy consumption by 2030.[16] One way to interpret this forecast is that DOE expects no meaningful improvement in the total energy efficiency of buildings over the next twenty years.

It is well within the ability of the building industry to defy this forecast. According to *The 2030 Blueprint* published by Architecture 2030,[17] 5 billion square feet of new buildings are built in the U.S. each year, 5 billion square feet of existing buildings are renovated, and about 1.75 billion square feet of existing buildings are demolished. Over the next 30 years, 75 percent of our building stock will be built new or renovated. How those buildings are built or renovated will determine whether the energy consumed by buildings can be meaningfully reduced. If we fail to do so, "fragmentation" of the industry will be no excuse; as noted in the *Blueprint,* significant reductions can be achieved with existing technologies in the existing building industry culture, "through proper design, i.e., building shape and orientation, natural heating and cooling, daylighting and ventilation strategies, proper shading, and straightforward, off-the-shelf building energy efficiency measures." These strategies do not require the development of new technology, the implementation of alternative technologies, or additional capital investments. Instead, they require an alignment of the building design with the surrounding physical environment, both natural and man-made.

Most commercial building owners pass on the cost of operating their buildings to their tenants and therefore have no direct incentive to increase the energy efficiency of their buildings. But commercial building lessees are becoming more and more sophisticated with respect to their total leasing costs and are increasingly demanding—as a matter of corporate policy—that the space they occupy have the least possible deleterious effects on the environment. The total operating cost, energy efficiency, and "greenness" of leased space are becoming increasingly significant factors in leasing rates. For building owners, minimizing their own costs at the expense of their tenants' operating costs is becoming less and less of a viable option.

CONVERTING INEFFICIENCY AND WASTE INTO PROFIT

In 2004, the Construction Industry Institute (CII) estimated that up to 57 percent of construction spending in our current business model is non-value-added effort or waste (see Figure 1.3).[18] With a U.S. market estimated at $1.288 trillion for 2008, over $600 billion in waste annually—if the CII estimate is accurate—is waiting to be recovered as profit by enterprising companies.

In any business enterprise, every dollar of unnecessary expense is a dollar of lost profit. Inefficiency and waste are unnecessary expenses that every business enterprise should ruthlessly eliminate. The historical fragmentation of the building industry is no longer the reasonable excuse for inefficiency and poor productivity that it has been for the last sixty years. Opportunities for improved efficiency and productivity are available now across the board and throughout the life cycle of buildings. Inefficiency and waste in any process are prime targets for greater profit. Achieving meaningful improvements in efficiency and productivity on every project and in every organization will depend upon access to reliable building information that can be created, exchanged, analyzed, modified, and updated throughout the useful lives of buildings. The obstacles to improved efficiency in the building industry are no longer primarily technical. Instead, it is the lack of reliable, timely information that makes efficient behavior and strategic decision-making difficult.

FIGURE 1.3
Construction/Manufacturing Waste Comparison Pie Chart. (Source: Eastman et al., *BIM Handbook*. © 2008 by John Wiley and Sons, Hoboken, NJ. Reprinted with permission.)

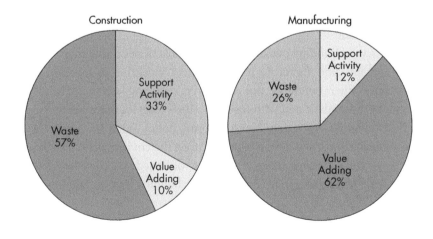

The potential for recycling of building materials is just one example of how accurate, readily available building information can foster strategic decision making, improve efficiency, and boost productivity. Currently, construction waste accounts for nearly 40 percent of the volume of material disposed in landfills.[19] Construction waste is material generated in both the putting up and the tearing down of buildings. A lot of those materials are reusable, either in their original or some altered form. The problem is that they are typically disposed in a heterogeneous pile that is then incorporated into a heterogeneous waste stream, from which it becomes uneconomical to extract them for recycling or reuse. In their original incarnation, however, they exist in a highly structured form; in a building, no one has any difficulty distinguishing ceiling tile from drywall from carpeting from cast-in-place concrete. During construction, neither is it difficult to distinguish building materials from the packaging in which they are delivered. These materials—for both construction and demolition, become "waste" only when they are mixed together, degraded, contaminated, or otherwise rendered unsuitable for their original purpose.

If the quantity and location of materials and packaging could be tracked efficiently and accurately in a building information model, or BIM, it would be much easier to plan strategically for their eventual reuse or disposal. Buildings scheduled for demolition or renovation in which the types and quantities of materials to be removed are known could be "deconstructed" instead of demolished, because the value of their embodied "raw materials" could be quantified. The market for these materials, and the cost and sequence of removing them, could be determined in advance. The potential exists not only for reducing the waste stream, but for dramatically reducing or even eliminating demolition costs, depending on the value of the raw materials that could be "mined" from a building. Spot markets have developed for copper pipe reclaimed from buildings scheduled for demolition, and the reclamation of antique heart pine and heavy timber hardwood from nineteenth-century industrial buildings (and river bottoms) is a mature industry. Both are strong indicators that large-scale material reclamation is a potentially viable business model for purely economic reasons, regardless of its environmental benefits. An entirely new segment of the building industry—deconstruction—is just waiting to be born, and the midwife is building information.

BENCHMARKING WASTE

The United States is home to one of the world's most effective and prosperous vehicle recycling industries. Today, 95 percent of cars retired from active use each year are processed for recycling, with about 75 percent of a car's material content (steel, aluminum, copper, and so on) eventually being recycled for raw materials use, including material that goes back into the manufacturing of new parts for new automobiles.[20] Putting these materials back into the supply chain increases raw material supply and lowers raw material costs across the board, to say nothing of the environmental benefits that accrue from not having to harvest and process additional raw material in the first place. This feedback loop occurs despite the fact that the original product has to be retrieved from the end user. Buildings, by virtue of being immobile, are much easier resource-recovery targets. It is only the lack of reliable information about their material contents that impedes the effective reuse of those materials and diminishes their value.

Material recycling is just one example of the business opportunities available—at the very end of the building lifecycle—to those with access to reliable building information. Similar opportunities for greater efficiency, productivity, and profitability are available to everyone—architects, engineers, constructors, and building owners—at every stage of the building lifecycle, especially in the very early stages of design.

IDENTIFYING BUSINESS OPPORTUNITIES

The core attribute of building information modeling that distinguishes it from the design technologies that preceded it is not three dimensional geometric modeling, but structured information—information that is organized, defined, and exchangeable. Unstructured information, by comparison, is difficult to identify, manage, or exchange. If you have to search for a needle in a haystack before you can sell the needle to a customer, the search for the needle might cost more than the needle's resale price; the needle becomes a cost center instead of a profit center. The lack of information about where the needle is stored renders the needle worthless; it is cheaper to go out and buy another needle from another, more efficient supplier, even if that means a reduction in mark-up and profit.

This scenario is repeated over and over again in the building industry throughout the lifecycle of buildings. The cost of gathering information about the actual equipment, materials, and components that make up a building can be so great as to render the physical artifacts worthless—or *worth less* than the cost of replacing them or subjecting them to thorough cost/benefit analysis.

In the design process, for example, it is often said that 80 percent of the cost of a building is determined in the first 20 percent of the design process.

This is another of those axioms that is insufficiently supported by statistical data, but industry professionals generally accept it as true. Once the major decisions are made about a structural system, a mechanical system, a cladding system, and so forth, the ability of the design team to control the cost of these components diminishes.

Design professionals—architects and engineers—rely on their experience and seasoned judgment to make critical decisions about the design and selection of major building components, materials, and systems in the early stages of design. Decisions are commonly made by applying rules-of-thumb or tried-and-true formulas that have not been rigorously tested against the specific requirements of the specific project. This is the best that can be hoped for in the current business climate. In the absence of reliable building information and the ability to share it easily among team members, the research needed to test different design scenarios and conduct rigorous comparative analyses of every plausible design option for every project would be cost prohibitive.

Substantially missing from the conventional building design process is the body of knowledge that has been delegated to constructors about the means and methods of construction. Without this knowledge, even the best design decisions by the most knowledgeable designers are made in a partial vacuum. In warfare, battlefield success up to the era of the Civil War often depended on the ability of a field commander to guess correctly the position of the enemy, based on the commander's knowledge and intuition of the opposing commander's judgment. Subsequent advances in aerial (and later, satellite) reconnaissance eliminated the guesswork. But for the design of buildings today, we still rely a great deal on this sort of intuitive judgment.

A second and even larger factor in building design is time. Time is money, and during the building design and construction process time has two parallel units of measurement: the total amount of "chargeable" time spent by design professionals to design and constructors to build the building (labor cost), and the "calendar time" between inception of a project and final occupancy of the facility. Chargeable time is easily quantified. The cost of calendar time, which can exceed chargeable time by a wide margin, is much harder to determine. It includes the cost of inflation, the nonperformance of physical and capital assets during the design and construction period, the cost of financing, and other "soft" costs such as legal and accounting fees, permits, and so forth.

With the clock (and money meter) continually ticking, design teams are under enormous pressure to make decisions quickly, even if the decisions are suboptimal. It does not take long before time-related costs exceed the potential savings of more-intensive design analyses. The challenge for designers and constructors is to reduce the cycle time and increase the value of conceptual design by leveraging richer sources of reliable information in the early stages of the design process. Building information modeling creates opportunities for scenario planning and rapid prototyping to determine optimum design solutions while potentially shortening the design and construction schedule.

The pattern of inefficiency repeats itself throughout the building life cycle, beginning with the warranty period and extending through operation and maintenance of the building until the end of its useful life. Information about components, materials, and systems is continually lost, and the cost of regathering that information quickly becomes cost-prohibitive. Facility managers, like design professionals, have no choice but to rely on their own experience and seasoned judgment, and their own rules-of-thumb or tried-and-true formulas. Staff changes within an organization result in further loss of specific, reliable knowledge about the facility. Inadequate maintenance—because reliable information about proper maintenance is not systematically available—leads to suboptimal equipment performance and premature equipment or system failure. The lack of documented maintenance history may itself compel premature replacement. If you don't know the exact age of a piece of equipment, its maintenance history, and its estimated useful life, its continued operation can become a gamble with life safety implications. Before long, the safe and prudent decision is to replace it. Worst of all, the loss of the original performance specifications may result in replacement with new components that do not meet those specifications. This scenario repeats itself again, and again, and again, to the frustration of all concerned. Expediency rules at every turn, simply because good information is not available for making better decisions.

By contrast, a database of structured information is a tangible asset that can enhance the value of a building. A true database of structured information—a building information model—also enables different parties to view the data from their own point of view. Computer-aided design (CAD) had little impact beyond the design and construction phase of buildings, because the data output—two dimensional, diagrammatic, pictorial representations of buildings—were of little value to facility managers, who view building information primarily in alphanumeric form. Most of the information created during the design and construction process that is of value to facility managers can only be found elsewhere and in scattered sources: in written construction specifications, warranty certificates, and operations and maintenance manuals. Any information contained in construction documents—CAD drawings and written specifications—that might be relevant to facility managers has to be extracted, yet another arduous, inefficient, needle-in-a-haystack task.

The core attribute of BIM—structured information—opens the door to easier and more effective *building information* transfer at every critical juncture of *building stewardship* transfer. The full potential of this is yet to be realized. The compilation of building information during the design and construction process, and the necessary electronic information exchange protocols, need to reach a certain level of maturity before useful facility management information can be conveyed routinely and easily to facility managers. But the structured nature of building information models provides the necessary infrastructure to facilitate this technological and business process development.

EMERGING BUSINESS STRATEGIES

The construction industry today is not unlike the sailing industry of a century ago. Skilled, seasoned mariners relied primarily on a method known as "dead reckoning" for coastal navigation. Complex calculations based on the inputs of wind speed, water current speed, and compass settings would produce—at best—a rough guess of the correct course to navigate. Dead reckoning was a poor substitute for the preferred method of coastal navigation: sighting prominent physical landmarks such as lighthouses, and either navigating toward them or away from them based on knowledge of coastal conditions as shown on nautical charts. The crude metrics of dead reckoning became most evident—and deadly—when they were most needed: when landmarks could not be seen due to weather conditions.

Despite the lack of reliable information, seasoned and experienced navigators managed to reach their intended destinations by the most expedient possible route most of the time. Long-range, low-frequency radio navigation (LORAN) systems eventually replaced dead reckoning in the mid-twentieth century, greatly increasing the accuracy of coastal navigation and dramatically reducing the risks of ocean shipping. In the late twentieth century, satellite-based global positioning systems (GPS) enabled mariners to determine their precise location anywhere on the globe with pinpoint accuracy. Mariners today are no more—and no less—skilled than their predecessors of a century ago. Like modern army field commanders, they simply have access to better information. Meanwhile, the life-or-death "data points" they formerly relied upon— lighthouses—have become artifacts of nostalgia with no useful function.

The real value of BIM to any organization—whether it is a design firm, construction firm, or building owner—lies in leveraging the structured information contained in a building information model to create value. The first step is a critical evaluation of the organization's core competencies and business objectives, followed by strategic deployment of appropriate technology to take the guesswork out of business decisions and shift the organization's output from routine, low-value-added tasks and services toward high-value-added tasks and services.

Moving from an unstructured to a structured information environment is neither cheap nor easy. Mariners skilled in dead reckoning were loath to trust the newfangled LORAN technology when it first appeared on the market after World War II. Many building industry professionals initially will be reluctant to substitute trust in the accuracy of structured information for their tried-and-true, rule-of-thumb methods for building design, construction, and operations. Confidence in the reliability of building information is a necessary prerequisite.

In order to make meaningful progress on this score, the industry needs to reach the level of information assurance that now prevails in the eBanking and eCommerce industries. The public overwhelmingly accepts that financial transactions executed at any ATM in the world, or purchase transactions executed online, will be executed properly. The occasional system failures, transaction errors, or outright thefts are regarded as aberrations, not fundamental defects. The building industry needs to develop the same level of confidence in building information and needs to provide its customers with the same level of information assurance. The only way to do so is to develop a profound understanding of the digital building information by engaging it, interacting with it, and exploiting it.

CHOOSING THE RIGHT TOOLS, DEPLOYING THE RIGHT TOOL SUITES

The selection of the most appropriate software solutions for individual firms is extremely important. Software should be selected for one reason and one reason only: to enhance the revenue-generating potential of the company. For a design firm, the selected software should enhance its ability to design; for a specialty consulting firm, its ability to perform iterative analyses; for a construction firm, its ability to build; for a building owner, its ability to manage and maintain its real property. In every case, software should enable every firm to do more with less. In all cases, software should enhance the ability of individual firms to communicate with other firms and exchange information reliably. If an investment in software does not increase productivity, streamline workflow, increase the quality of goods and services produced, reduce operating costs, and increase profits, then it does not meet the definition of a technological advancement and should not be deployed.

Business leaders have a tendency to evaluate technology on the basis of its acquisition cost rather than its full implementation cost and full revenue-generating potential. There is a great deal more to a strategic technology plan than software licensing and training, which are often viewed and managed as overhead expenses to be controlled, rather than as components of a larger strategic investment that should produce a measurable financial return. The larger, often hidden investment is in the education (as opposed to mere training) that will enable an entire organization to change its business culture, and in the resulting reform of core business processes to achieve greater productivity than can be achieved by simply automating existing processes(see Figure 1.4).

The result of the cost-based view of technology is that most software is grossly underutilized, either because the software is poorly matched to the firm's business needs or the firm fails to fully exploit the technical capabilities of the software it already has. This is so common because so few firms conduct the rigorous, critical assessments of their software applications' functional capabilities that would determine how well those capabilities align with their business goals.

BIM authoring tools—the large, robust applications that are used to create and compile most of the information contained in a building information model—are often perceived as costly to purchase and deploy. The failure to perceive BIM software as an investment is compounded by the failure to recognize that the cost of the software is only a small fraction of the total investment in BIM. The problem is particularly acute in design firms, which tend to make

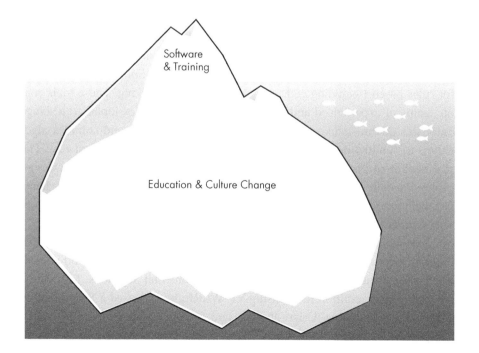

FIGURE 1.4
The Hidden Costs and Benefits of BIM

Software & Training

Education & Culture Change

little effort to measure the return on their investment in BIM. As a result, design firm leaders' perception of the impact of BIM on their firms is often grossly inaccurate, which makes it impossible for them to make the correct strategic decisions.

Though BIM is relatively new technology, distinct categories of BIM software have emerged. Authoring tools—which include Autodesk Revit, Bentley Architecture, Graphisoft ArchiCAD and Nemetschek Vectorworks, among others—are optimized for building design. Each has distinct characteristics or attributes that make it suitable for particular types of design firms serving particular markets. Constructors need to look beyond these tools to other software applications—such as dProfiler by Beck Technology and Constructor by VICO Software—that have specific functionality for construction cost estimating, constructability analysis, and construction sequencing. Building owners and facility managers need to look even further for tools that are suitable for facility management, operations, maintenance, and real asset management, such as Archibus, ArchiFM, Drawbase, Facilivue by Omegavue, FM:Interact and FM:Space by FM:Systems, NETFacilities, and the MARS Facility Cost Forecast System by Whitestone Research.

For design firms, BIM authoring tools are merely the first and most powerful weapon in the design arsenal. Audit and analysis tools, typically far less expensive than authoring tools, have a far greater direct payback. They are easier to learn because they typically are designed to do one thing very well. Design firms are now using such tools for clash detection, energy analysis, sustainable design analysis, code compliance, and construction cost estimating. Applications in this category include Autodesk Navisworks for clash detection; Ecotect and IES VE-Ware for energy analysis; and Solibri Model Checker for rules-based (including code compliance) model checking. These tools enable teams of experienced and knowledgeable design professionals to leverage databases of statistical, technical, or financial information and complex algorithms to conduct detailed analyses of specific designs at a marginal cost. The real benefit of audit and analysis tools, however, is that they enable design professionals to enhance and leverage the value of the very thing that clients pay them for: their professional judgment. For design firms, tools in this category are essential to increasing efficiency, productivity, profit, and value.

BIM audit and analysis tools continue to proliferate because the volume of building information available for analysis is continually growing, and with it, the market for analysis tools. Integrated Environmental Solutions, Ltd. (IES), for example, in addition to releasing VE-WARE, has released a Sustainability Toolkit for sustainable design analysis and a LEED toolkit for LEED credit compliance analysis, among others. These toolkits can only be used to analyze building information models created in other applications: BIM authoring tools. So the market for specialized audit and analysis tools depends entirely on the growth of the BIM authoring software market. Many more such specialized applications will emerge as the volume of structured building information grows and software developers capitalize on the market opportunities for software that facilitates the analysis of structured building information. Continually surveying the BIM software market and evaluating these specialized tools is an essential component of a design firm's strategic business plan.

For the very early stages of projects, architects, planners, developers, and government agencies should consider predesign tools that can be exploited to accelerate, automate, or streamline tedious and expensive information-gathering or decision-making processes. Tools in this category include the Onuma Planning System by Onuma, Inc., and Affinity by Trelligence, Inc., both of which are specifically designed for the scenario planning, master planning, and programming stages of projects. While these tools can be deployed by anyone, they are a natural fit for design firms that have developed core competencies in planning and programming services, and for any government agency engaged in land planning and the related consensus decision-making processes.

Firms throughout the industry should look beyond BIM for technologies that can further enhance the benefits of BIM. A building owner or a design firm specializing in historic preservation or the adaptive reuse of existing buildings, for example, should consider deploying the laser-scanning technology of companies such as Quantapoint or Intelisum to document existing conditions. These technologies can dramatically reduce documentation time and cost while significantly improving the quality and detail of the information gathered. The integration of these applications with BIM applications is continually improving.

THE BIM VALUE PROPOSITION

As part of their "due diligence" in assessing the deployment of BIM in their businesses, design firm leaders often ask how they can pass on the added cost of developing robust BIM models to their clients. Construction firm leaders, on the other hand, ask how they can exploit the technology to reduce their own project-related expenses and increase their profits. Several early BIM adopters in the construction industry are reporting astounding results from their BIM implementations. Design firm leaders would do well to view the technology through the same lens. The key to leveraging BIM technology to increase profitability is not raising fees but rather reducing cycle time and increasing value. Clients will only pay more for something if they perceive that it has greater value, and the value of BIM models to clients, for now, can be difficult to demonstrate as a hypothetical future benefit. Asking clients to pay more without delivering more is a dead end. The value of BIM must first be proven. By focusing on increasing the efficiency of their own internal operations and the productivity of their own design teams, design firms can demonstrate the value of BIM to their clients while increasing their own profitability. By shifting their own perception of their services from cost-based to value-based, design professionals also may succeed in shifting their clients' perceptions as well, enabling them to earn higher profits while reducing their clients' costs.

PROCESS ENGINEERING

To some extent, building industry business leaders will need to recognize the seemingly perverse effect of efficiency and productivity on revenue. Improvements in efficiency and productivity resulting from BIM will cause the unit cost of design and construction services to go down while the quality, value, and profitability of those services will go up, because lower costs will increase total market size. This is an axiom of economics that is taken for granted in other industries such as automotive or computers, but is more difficult for building industry professionals to embrace. The unit cost of computing power of personal computers is only a fraction of the cost of the first computers introduced in the early 1980s, but the computer industry remains highly profitable. Toyota sells cars today of much higher quality and value but at a lower relative price point (after adjusting for inflation) than it did thirty years ago, but continues to enjoy record profits. Henry Ford did the same thing a century ago when he introduced the Model T. (Ford went so far as to pay his workers a substantial premium over then-prevailing wages, simply to increase the market size for his product.)

The same inverse relationship between unit price and profit has always been and will continue to be the hallmark of innovation. An entire professional discipline—process engineering—is devoted to it. Whenever the amount of time—or the unit of labor—needed to complete a task can be reduced, the efficiency and profitability for completing that task increases. Whether the additional profit accrues to the producer or is passed on to customers is partly a business decision and partly a matter of market dynamics. Individual businesses may not be able to control the market forces that determine the market value of their services, but every business is fully in control of how well it maximizes the value of its core competencies and how efficiently it delivers those services. That is why Toyota has been able to exploit innovation to earn record profits while U.S. automakers lose money in the same market.

Some increased efficiency can be achieved through incremental improvements in workflow, but the greater gains will come from transformational changes in business processes. If you're a wholesaler of sewing needles, you might increase your profit by 10 percent by finding a cheaper supplier, but you could increase your profit 400 percent by not storing your needle inventory in haystacks—and the building industry is full of haystacks.

The building industry did not invent such business concepts as invoicing, which accelerated business transactions by separating the delivery of goods and services from the related financial transaction, or just-in-time delivery, which dramatically reduced inventory costs in manufacturing, wholesaling, and even retailing. When these business practices were adopted by the building industry, they did not inaugurate much introspection and analysis. It appears that BIM, however, is causing the building industry to cast a critical eye on other business processes and ask how these can be reorganized in a structured way, whether the process is design, construction, marketing, communication, project management, professional development, or financial management. How much time does your highly paid staff spend managing e-mail? Processing submittals and RFIs? How frequently does your accounting department deliver project financial information to your project managers? Are your project managers able to act on the information provided? How effectively are you able to assess staff performance? How much do these nonchargeable (and therefore no-value) tasks cost your firm?

These are questions that are directly related to the availability of structured *transactional* information that is beyond the immediate realm of any individual project and, for the most part, beyond the realm of BIM technology. These are enterprise-wide issues of business information flow that can only be addressed by applying a strategic approach to enterprise workflow. Firms that focus solely on BIM may realize gains in project delivery only to be undermined by rising operating costs in other areas.

The disciplines of process engineering and supply chain management are largely absent from the building industry, but this will inevitably change. The most enterprising firms in the industry can be expected to recognize the void and begin acquiring this expertise and applying it to their own enterprise workflows, linking suites of specialized tools in customized end-to-end solutions that will significantly increase the value of their services. Organizations that fail to innovate or lag behind will find themselves at a significant disadvantage in the marketplace. As the Toyota and Ford example shows, innovation can come from any quarter and disrupt what seem like safe and secure markets.

 THINKING LIKE AN OWNER

Design and construction industry professionals are often caught looking through the wrong end of the telescope, focusing on improvements in the design and construction of buildings rather than on the impact that design and construction has on the total life cycle costs and operations of a building, or the total environmental impact of design and construction decisions.

Buildings are built to serve a purpose. The true value of a building to an owner is the product or service produced by the people who occupy it, or the fulfillment of the mission that the building is intended to shelter. The cost of the facility itself is typically a small fraction of the cost of the operations or the value of the activities that it houses. Over the typical twenty-year life of a commercial building, for example, 90 percent of its total facility cost can be attributed to the payroll cost of the people who occupy it—a 9:1 ratio. The remaining 10 percent is evenly divided between the original construction cost and twenty years of operations and maintenance (including energy consumption).[21] In manufacturing, the ratio of total production cost to facility cost can be 100:1 or greater. It is easy to understand, then, why the owners of commercial or manufacturing facilities devote scant attention to their buildings, regarding them as little more than containers for the revenue-generating operations that take place within them. Though the metrics may vary, the same principles apply whether the building is a factory, office building, hotel, apartment building, convention center, dormitory, shopping mall, airport terminal, hospital, research laboratory, sports stadium, theater, church, or single-family home.

From an owner's perspective, however, anything that interferes with or diminishes the revenue-generating potential of the operations within a building is detrimental to the owner's business interests. If a $10 million business operation cannot take place because a $250,000 building is not ready for occupancy or is unavailable because $100,000 was spent on a roof that chronically leaks, the relatively small facility-cost investment suddenly takes on outsized importance. NASA spends so much money on every component of its habitable nonplanetary assets not because it wants to build best-of-class habitable environments, but because the consequences of component failure are so high.

Other pressures can compel short-term thinking. At the project inception stage, building owners face acute cash-flow challenges. Because a building cannot generate revenue until it is occupied, construction financing can be difficult and expensive to obtain. The only assurance that the lender has of recovering its investment is for the building to be completed and begin generating revenue or serve its intended purpose; a building that is 95 percent complete is a liability, not an asset. So before a building is built, the fact that design and construction costs are such a small fraction of total life cycle cost doesn't matter very much. The dollar amount that needs to be financed and the amount of time it will take to complete the building are the owner's and the lender's paramount concerns. It is little wonder, then, that owners strive to limit both construction cost and construction time regardless of the long-term consequences.

Building industry professionals often overlook these mission-critical financial considerations. It may come as a shock to design and construction professionals, but from an owner's point of view, the entire time between the moment a decision is made to erect a building and the moment the building is ready for occupancy is an obstacle to achieving the owner's business mission. For design and construction firms that are truly client-focused, the primary goal should be to complete the building and get it into service as quickly as possible. For long-term consequences to be given greater consideration, design and construction time cannot be lengthened.

BUILDING PERFORMANCE METRICS

While the building owner's focus on minimizing initial construction costs is understandable, it is grossly shortsighted. The cost of operating and maintaining a building—and the efficiency and productivity of the activities that take place within it—are all directly related to the quality of its original design and construction. The quality of the physical environment has a direct effect on the health and well-being of the people who occupy it, which in turn directly affects their productivity, as numerous studies on workplace productivity have shown.

Just as structured building information allows architects and engineers to incorporate environmental considerations and life cycle operating and maintenance costs into the design process cost-effectively, building information modeling will enable workplace productivity factors to be taken into account in equally methodical ways based on reliable statistical workplace performance data. This will dramatically alter the value proposition of buildings and the business environment with respect to cost/benefit analyses of design alternatives.

NEW METRICS FOR REAL PROPERTY VALUATION

Reliable, accurate life cycle building information will change the metrics for real property valuation. Currently, buildings are valued largely on the basis of their physical attributes such as square footage, location, and rough measures of "quality" such as Class A, B, or C office space, "affordable" versus "luxury" housing, and three, four, or five-star hotels. While the latter three metrics bear a relationship to a building's revenue-generating potential, they are loosely based on physical attributes grounded in the original design and construction quality and cost. When the true life cycle costs and benefits of buildings can be reliably quantified, these factors will be monetized and reflected in the value of buildings alongside physical factors, which will then account for a proportionally smaller fraction of a building's total value.

Historically, it has been all but impossible for building owners to measure the relative life cycle costs and benefits of design and construction decisions because, once again, the cost of the research and analysis needed to develop reliable forecasts has been prohibitive. Building information modeling provides an information infrastructure that will allow architects, engineers, constructors, and owners to assess multiple life cycle factors in the early stages of design collaboratively, including energy consumption, total life cycle cost of materials, equipment and systems, and workplace productivity. Project teams will also be able to assess the total environmental impact of material, product, and equipment selections, and not just their impact on energy consumption during building operations or their effect on the environmental health of the building's occupants.

ENDNOTES

1. David Malin Roodman and Nicholas Lenssen, Worldwatch Paper 124: "A Building Revolution: How Ecology and Health Concerns Are Transforming Construction" (Worldwatch Institute, 1995).

2. Grecia Matos and Lorie Wagner, "Consumption of Materials in the United States, 1900–1995," *Annual Review of Energy and the Environment,* no. 23 (1998).

3. Ibid.

4. Population Division of the Department of Economic and Social Affairs, "World Population Prospects: The 2004 Revision" (United Nations Secretariat, 2005).

5. Ibid.

6. "U.S. Population from 1900," Wendell Cox Consultancy, http://www.demographia.com/db-uspop1900.htm.

7. Robert N. Lubowski et al., "Major Uses of Land in the United States, 2002," ed. United States Department of Agriculture (Washington, DC: Economic Research Service, 2006).

8. "Agricultural Productivity in the United States," United States Department of Agriculture, Economic Research Service, http://www.ers.usda.gov/data/ agproductivity/.

9. Keith O. Fuglie, James M. MacDonald, and Eldon Ball, "Productivity Growth in U.S. Agriculture," ed. United States Department of Agriculture (Washington, DC: Economic Research Service, 2007).

10. Ibid.

11. Bureau of Labor Statistics, "Industry at a Glance: NCAIS 23: Construction," United States Department of Labor, http://www.bls.gov/iag/ construction.htm.

12. Paul Teicholz, "Labor Productivity Declines in the Construction Industry: Causes and Remedies," AECBytes, April 14, 2004, http://www.aecbytes. com/viewpoint/2004/issue_4.html.

13. Preston H. Haskell, "Construction Industry Productivity: Its History and Future Direction," *The Haskell Company White Papers* (Jacksonville, FL: Haskell Company, 2004).

14. Ibid.

15. Office of Integrated Analysis and Forecasting, *Annual Energy Outlook 2007* (Washington, DC: Energy Information Administration, U.S. Department of Energy, 2007). Consumption figures are for residential and commercial buildings only. Energy consumed by industrial buildings is not included.

16. Ibid.

17. Edward Mazria and Kristina Kershner, *The 2030 Blueprint: Solving Climate Change Saves Billions* (Santa Fe, NM: Architecture 2030, 2008).

18. James E. Diekmann et al., *Application of Lean Manufacturing Principles to Construction* (Austin, Texas: Construction Industry Institute, 2004).

19. U.S. Green Building Council, *Green Building Facts* (Washington, DC: 2008).

20. Alliance of Automobile Manufacturers, "Recycling Vehicles," The Alliance of Automobile Manufacturers, http://www.autoalliance.org/environment/ recycling.php.

21. Federal Facilities Council, "Federal Facilities Beyond the 1990s: Ensuring Quality in an Era of Limited Resources" (Washington, DC, 1997).

CHAPTER 2

BIM Implementation Strategies

Even if you're on the right track, you'll get run over if you just sit there.

—Will Rogers

Implementing building information modeling is much more of a business decision than a technical one. BIM is an enabling technology with the potential for improving communication among business partners, improving the quality of information available for decision making, improving the quality of services delivered, reducing cycle time, and reducing cost at every stage in the life cycle of a building. But while it opens the door to these possibilities, it does not make them happen. The technology must be deployed as part of a comprehensive business strategy in order to be successful. Many business processes and workflows must change to take full advantage of the technology.

Maintaining or enhancing one's competitiveness in the marketplace, or streamlining one's business operations, are among the reasons most commonly cited by business leaders for implementing BIM. These are perfectly valid business reasons, but when questioned, few business leaders can articulate a coherent strategy for how building information modeling will enhance their competitiveness or streamline their operations. Many fall back on a desire not to be perceived as lagging behind a growing and inevitable trend.

For many business owners and senior managers throughout the building industry, the key decision in their BIM implementation strategy is which software application to buy, and the key criterion for selection is "what everyone else is using." This is followed by decisions about the number of software licenses to purchase and the number of staff members to send to training. Too often, these three things define an organization's entire BIM implementation strategy. This is a legacy of the CAD era, in which a deep-seated aversion to risk and inexperience with the novel technology made following the herd appealing. In building design as in finance, no one ever got fired for following the market. But just as a conservative investment strategy yields only marginal returns, the conservative technology strategy of design firms in the CAD era yielded only marginal gains in innovation.

These are troubling business phenomena. In any realm of decision making other than technology, the same business leaders would consider the allocation of significant capital resources in the absence of a clearly defined business strategy highly irresponsible. They would not be likely to put the selection of a single product or service at the center of a comprehensive business strategy, nor would the popularity of a product among their competitors figure highly among their selection criteria. Successful business leaders develop business strategies to *distinguish* their companies from their competition. They examine their business needs and select products and services that meet those needs. They conduct cost/benefit analyses to assure themselves that the investment will result in increased revenue and profit. They establish performance metrics and closely monitor the return on their investment to determine whether it is yielding the forecasted results. They use the knowledge gained from measuring performance to adjust their strategy or to make better investment decisions in the future.

All too often in the building industry, investments in information technology seem to get a "pass" on this type of due diligence. This is a legacy of a business culture, now a generation old, that viewed technology in much the same way scientists view research in basic science—the pursuit of knowledge for its own sake, with no expectation of a practical result—but without the accompanying rigor of the scientific method, which requires that assumptions be documented and results compared to those assumptions.

LEAVING THE CAD ERA BEHIND

Because of our lack of due diligence and the absence of statistical data about the impact of technology on the building industry, we know frighteningly little about the impact that information technology has had on the industry over the last twenty or thirty years. This undisciplined approach to technology deployment in the building industry must change, if for no other reason than that we have failed to achieve the productivity gains realized in other industries over the same period. And it is becoming increasingly clear that the companies harvesting the most significant gains from their implementation of BIM are those that have exercised due diligence in their BIM implementation strategies. If the gains these firms have achieved are to be realized by the industry as a whole, companies throughout the industry will have to make decisions about BIM with the same rigor and discipline.

BIM is a technology that affects business processes beyond drafting and well beyond the organizational boundaries of design firms. The full potential of the technology cannot be realized with a narrow focus on the technology. The cost of software and training, while not insignificant, is incidental when compared to the potential impact on your organization's profitability resulting from other aspects of BIM.

Software is a consumable commodity, not a capital investment. It is of value only to the extent that it enables your organization to fulfill its mission. We can safely anticipate that software technology will change. The cost of training is an ongoing operational expense, not a strategic investment. Within the scope of a comprehensive BIM implementation strategy, software selection and training decisions must be made in the context of broader business objectives.

For a BIM implementation strategy to be fully effective, software training must be preceded by, or at least accompanied by, education. Training teaches people how to do. Education teaches people how to think. Employees trained to use a BIM application will learn how to perform tasks, not how to improve or change business processes, which only business leaders can do. An effective (and documented) BIM implementation strategy is necessary to provide the framework for an effective BIM training program.

A SYSTEMS APPROACH TO BIM IMPLEMENTATION

Business owners need to be able to perceive tangible benefits to changing internal business processes before they will make the investment to implement those changes. They also need to be able to recognize tangible benefits to changing the nature of their business relationships with business partners and clients. Whenever business owners are able to perceive such benefits clearly, BIM tends to foster rapid change, and institutional, legal, or cultural obstacles are easily overcome. On the other hand, when business owners are unable to connect BIM implementation to clear business goals, change tends to occur slowly or not at all, and the obstacles seem insurmountable. The most effective BIM implementation strategies are those based on a thoughtful review of an organization's business processes and workflow, both internally and externally. The focus is not on how to adapt the workflow to suit the technology, but rather on how to exploit the technology to improve the workflow.

The key to any successful BIM implementation is recognizing that an organization's internal business processes—whether it is a design, construction, or property ownership enterprise—are part of a *system,* and that building information created by anyone in the system is of potential value to anyone else in the system. This does not mean, as is so often asserted, that one party must assume the responsibility or bear the risk and cost of creating, compiling, or maintaining a comprehensive building information model for the benefit of someone else. In order for building information models to be sustainable throughout the life cycle of a building, they must be created and maintained on a sustainable business model, with a clear value proposition between the creators and hosts of the model and the beneficiaries of the information contained within it.

AVOIDING IDEOLOGICAL PITFALLS

A systems approach to building information modeling should not be confused with the notion of a single building information model. Implementing BIM does not mean that all of the information about a building must be compiled into a single data file, reside in a single physical location, or be maintained by a single business entity throughout the life cycle of a building. The notion of a comprehensive life cycle building information model—while conceptually appealing—is problematic from a business point of view. Often cited as one of the primary goals of a BIM implementation strategy, the single building model is beyond the reach of any end user today for the same reason that it has been out of reach for thirty years: neither the technology nor the market conditions needed to support it exist. To the best of our knowledge, not one viable comprehensive building information model residing within a single data file has ever been created.

No commercially available software application or technology platform is capable of containing all of the information created about a building throughout its useful life and making it accessible to appropriate stakeholders in real time on demand. More significantly, *none is in development*. The unmistakable trend in building information modeling software development is toward distributed building information models created by highly specialized software tools *that are designed to work together*. A number of factors may have contributed to this trend:

- The entire building life cycle of business processes and workflows is too complex to be modeled effectively within a single software application.
- Business processes and workflows vary too much across the industry and across the building life cycle to fit neatly within a single workflow paradigm.
- Working within a single building model environment requires too great a change of existing information-management infrastructure and business processes to support viable migration paths from existing workflows to new ones.
- The cost and technical challenges of developing a software application capable of meeting the needs of all users throughout the life cycle of a building are prohibitive.

Consider, for example, the specific use case of an architect and a structural engineer collaborating on the design of a building. A single software application that included all of the functionality needed for both architectural and structural design would be extremely unwieldy. Only a portion of the available functionality would be of use to either the architect or the engineer, but each would be burdened with a more complex user interface. Neither would be willing to pay for the functionality that neither is likely to use. An additional layer of complexity would have to be added to allow each party to maintain responsible control over the information that each party creates. The added complexity of the user interface and the increased IT burden of managing access to the data by multiple parties would likely erode any efficiency gained from the single building model environment. Complexity, cost, and functional inefficiency increase exponentially as other disciplines are added to the mix.

While statistical data about BIM implementation of any type is hard to come by, the case studies of "successful" BIM implementation that have emerged thus far reveal that the data for a typical BIM project is a compilation of distributed models created and analyzed using a suite of specialized BIM tools. The paradigm of "standardizing" on a particular BIM application or platform is becoming less and less important.

ALIGNING A BIM IMPLEMENTATION STRATEGY WITH TECHNOLOGY TRENDS

It is important for business leaders to understand and adapt their BIM implementation strategies with the evolving state of the available technology. In lieu of developing tools to create and sustain a single building information model, software developers are creating tools that allow each player in the building life cycle—particularly in the design and construction phases—to work within their own modeling environments and periodically combine file-based models for collaborative work or comparative analysis. This growing trend in BIM-related software development is now firmly established and can be expected to continue, for all of the reasons cited above.

It is also important for business leaders to recognize that the building industry is only in the very early years of an era of unprecedented innovation and experimentation that is only partly driven by technology. We are witnessing the emergence of many different ideas and technologies, some of which will work better than others. If this new culture of innovation and knowledge sharing can be sustained long enough, the most useful technologies will have time to mature and the best industry practices will spread rapidly.

The failure of the single building information model concept to gain traction, for example, is not necessarily a bad thing. Its full implementation would have required the wholesale disruption of existing business practices, processes, organizational structures, contractual relationships, and even individual work habits. Any technology that requires such a complete break with the status quo has a high probability of failure, regardless of its merits. It's simply unrealistic to expect that a large and highly fragmented industry can adapt to such wholesale change on so many fronts all at once. In hindsight, it is a good thing that the industry did not lock onto this entirely new business paradigm based on an entirely new technology without having had the opportunity to test and adapt it under real-world conditions. Whether by chance or intention, the industry has managed to sidestep the early ideological goal while still advancing the development and implementation of the underlying core technology.

The emerging distributed building information model paradigm allows for a more flexible and orderly integration of new technology without requiring an immediate and wholesale reordering of our entire business culture. It allows business partners to test different business practices and workflows, gain insight from their experiences, and modify their approach in a continuous cycle of innovation. It allows individual business owners to adapt their internal business practices, workflows, and technology at their own pace. Across the industry, it allows for a great deal of experimentation to take place and for a variety of business models to emerge to suit particular markets or individual circumstances, enabling both technology and business practices to develop organically. Finally, the distributed information model paradigm fosters greater market demand for interoperability—the seamless, reliable exchange of digital data—which in turn creates the market conditions for a greater array of specialized software tools.

ASSESSING FUNDAMENTAL RISKS

Though innovation is fraught with risk, the risk of implementing BIM technology is far lower than the risk of implementing CAD technology a generation ago, because it is much easier to align available BIM technologies with an organization's internal business processes and core competencies and measure the results. The transition from paper to CAD was largely a leap of faith—even in hindsight, the return on the investment is almost impossible to measure. Early adopters of BIM technology, however, are finding the benefits of BIM much easier to quantify and are realizing very substantial early gains. The rapid payback lowers risk, which fosters still greater innovation. The sooner an organization can recoup its investment on a particular BIM technology, the sooner it is free to explore other technologies. The days of being "locked into" one software tool or one software platform are over. Even the most risk-averse business leaders can comfortably exploit BIM technology to their competitive business advantage.

FOSTERING A CULTURE OF INFORMATION STEWARDSHIP

The trend in BIM technology suggests that the most viable and flexible business strategy to BIM implementation is one that emphasizes the value of information exchange to support business processes *(modeling)* over the artifact that results from those processes *(the model)*. Most participants in the building life cycle—even very experienced and knowledgeable professionals—have only a rudimentary understanding of the business activities that precede and follow their own. This is unlikely to change. We simply can't expect everyone involved in the life cycle of a building to know everything about that building, including its past and its future. We can, however, expect all those involved to develop a better appreciation of how their activities fit into and affect others throughout the life cycle.

In biological ecosystems, the many organisms that make up an ecosystem have little or no knowledge of how their behavior affects other organisms, and no understanding at all of how the entire ecosystem functions. Yet ecosystems as a whole can exhibit a very high degree of complexity, efficiency, and even apparent intelligence. Biological organisms interact with one another largely through their environment; the behavior of one organism has consequences that affect the behavior of other organisms, and the "work product" of one organism becomes a "found resource" for others. This is a form of information exchange, even though there may be no direct or conscious communication between organisms—each organism behaves autonomously in response to purely external stimuli.

The level of consciousness that we need in the building industry is only slightly higher, requiring no more than a general awareness that information created by one person may be useful to another. It isn't necessary, or even practical, to expect everyone in the building industry to understand everyone else's business processes and to anticipate just how the information they create might be used, when, and by whom. A strategy that depends on such profound understanding of the entire building life cycle by all participants in the life cycle is doomed to fail. What is important, rather, is that anyone involved in any part of the lifecycle of a building—from the geotechnical engineer analyzing a building site to the renovation or demolition contractor—recognize that the tasks they perform and the information they create are a small part of a very long sequence or cycle of tasks. Anyone can readily understand and appreciate that any building information they create might be of value to someone else for some other purpose, even if they have no idea exactly how, when, why, or by whom. Systems-minded building industry professionals regard the information they create with an attitude of *stewardship* rather than *ownership*. They are mindful that their possession of the information is temporary and that it is of potential value to someone else after it is no longer useful to them. They organize, compile, and maintain information in the most structured, integrated, and accessible manner possible. They view information as a tangible asset and a living resource.

Biological ecosystems are useful as a metaphor in yet one more way with significant implications for addressing issues of liability in the building industry: each organism is solely responsible for how it uses the resources it finds in its environment. Every available resource in the environment is accepted "as is" by all organisms, and is evaluated solely for its usefulness. Its original source is irrelevant. The organism that may have left that resource behind is not held accountable for it.

Cultivating a business environment of information stewardship is possible without disrupting any existing business processes or operations, assuming any new risks, or changing any existing business relationships. It can generate benefits that are purely internal to an organization. It can begin well before a company begins using any BIM technology. It is fundamentally about getting one's own house in order. An attitude of *information stewardship* manifests itself in effective *information management,* a necessary prerequisite to effective *information modeling.* As more and more industry professionals gain a greater understanding of the value of building information created throughout the life cycle of a building—and learn to manage their own information accordingly—more and more will be able and willing to engage in value-added information exchange.

MANAGING CULTURE CHANGE

For a BIM implementation strategy to succeed, it must be accompanied by a corresponding cultural transformation strategy. Cultural transformation is a greater challenge to the industry than any technological transformation resulting from BIM. It will require that building industry business partners regard one another differently than they do today—as true partners and collaborators with a mutual interest in a successful outcome, rather than as adversaries and potential future litigants. It will require that the industry reach beyond technology and business practices to alter the prevailing legal framework, particularly with respect to dispute resolution. Some newly emerging model contractual agreements contain the novel provision that the parties explicitly agree not to sue each other. Instead, the parties agree to work together to identify problems and correct them. The potential impact of these changes on the way we do business is simply enormous. The amount of time now spent by various team members documenting their own actions as a bulwark against possible future legal action—a no-value task, as far as the project itself is concerned—can be shifted toward completing the project in the most expeditious manner possible.

A greater climate of trust among business partners is frequently cited as a feature of the new business climate for integrated project delivery (IPD) using building information modeling. While the intensive collaboration inherent in IPD does indeed heighten the value of trust in business relationships—which makes the careful selection of business partners far more important—it would be naïve to think that complex business relationships can be built entirely on trust or that project team members always will be able to pick and choose the other members of the team. There will always be project teams made up of team members who have never worked together or do not know each other. There will always be project team members who turn out to be unscrupulous or incompetent, who fail to act in good faith, or who seek to protect their own interests at the expense of others.

In other sectors of the economy, an environment of trust is strongly supported by "trust but verify" mechanisms. The entire worldwide banking system depends heavily on trust. Not a single bank is capable of surviving the sudden demand of its depositors for all of their assets, because the vast majority of those assets are loaned out to others to generate revenue for the bank. When a bank fails, there may be little difference between its balance sheet and the balance sheet of a competing, "healthy" bank. A principal cause of bank failure is a loss of trust.

The quixotic element of trust that holds up the entire banking system is supported by robust banking regulation and partial guarantees in the form of government-backed deposit insurance. Though the analogy to banking is imperfect, business leaders in the building industry need to employ comparable trust-building mechanisms to bolster the trust between business partners. The new business culture then becomes a self-sustaining ecosystem that continuously seeks equilibrium, with players who are unable to meet the industry's high standards for trust and quality performance routinely getting squeezed out.

USING TECHNOLOGY TO BUILD TRUST AND MITIGATE RISK

In the existing building industry business climate—particularly in the building design and construction stages—the transfer of information from one party to another poses considerable risk for the author of the information. The sending party may be held accountable for the quality, completeness, and accuracy of any information they transmit, regardless of their degree of responsible control. At the opposite end of the risk scale is "no fault" information transfer. In a biological ecosystem, for example, each organism accepts the outcome of another organism's behavior as a "found" natural resource; the receiving organism is solely responsible and liable for how it uses the resource.

Between these two extremes, the Internet is a useful model for the future building industry information stewardship culture. One can find a great deal of information on the Internet, but the quality of the available information varies widely. As information consumers, each of us is responsible for validating the "found" information and assessing the risk of using it. Validation consists of two parts: determining whether the information is from a trusted source and confirming the integrity of the information itself. Integrity is assured by comparing the content against a reliable database and testing the digital data format to ensure that it is not corrupted. Some of the BIM audit and analysis tools discussed in Chapter 1 can be characterized as "data validation" tools. We can expect to see an increase in the number of offerings in this software market segment as the volume of BIM data available for analysis grows.

The ability of information recipients to validate data will shift responsibility for data integrity from authors only to both authors and recipients, while lowering the risk for both. Design professionals will be able to conduct more rigorous analyses of their designs to minimize or eliminate errors and omissions, while building owners and constructors—who will have the same access to clash detection and other data validation tools—will be held accountable for detecting errors and omissions before they result in a financial loss. This is one way in which the technology will help shift the current adversarial business climate toward a more collaborative one, simply by improving the quality of building information available.

The current reflexive response to errors or problems—to identify the responsible party and assign blame—will shift toward an environment where project team members will work collaboratively to identify mistakes early and correct them promptly. Economic imperatives, not altruism, will drive this transformation. Getting the project completed well as quickly as possible will become more important than pointing fingers and collecting damages.

A culture of information stewardship and frequent information exchange also results in much greater transparency. When every team member has access to the same information in real time, it becomes clear to everyone who is responsible for what, which team members are meeting their obligations, and where bottlenecks are occurring and why. It will become far more difficult for team members skilled in generating mountains of obfuscatory, finger-pointing paperwork to gain an edge over another team member. In a collaborative environment of information exchange, it is not just the emperor but his entire court that has no clothes. BIM, information stewardship, and information management are three interdependent components of a single integrated technology and business process.

MAINTAINING DATA EXCHANGE CAPABILITIES

An initial step toward greater information stewardship in any organization is an assessment of existing information storage, retrieval, and exchange capabilities. The more flexibly information can be exchanged, the greater the likelihood that it can be preserved in a useful form for the long term. A whole range of data exchange and storage options already exist. For example, many building design

Table 2.1   IFC Software Compliance Chart.

Application Name	Release
Active3D	4.2
Allplan	2006.2 / 2008
ArchiCad	10 / 11
AutoCAD Architecture 2008	2008 SP1
Bentley Architecture	8.5.3 / 8.9.4
Bentley Building Electrical Systems	8.5.3 / 8.9.4
Bentley Building Mechanical Systems	8.5.3 / 8.9.4
Bentley Structural	8.5.3 / 8.9.4
Bocad-3d	
DDS IfcViewer	
DDS-CAD Building	
DDS-CAD Electrical	
DDS-CAD HVAC	
DDS-CAD MEP	6.4
DDS-CAD Plumbing	
EDMdeveloperSeat™ Basic	
EDMdeveloperSeat™ Professional	
EDMmodelConverter™	
EDMmodelMigrator™	
EDMServer™	4.5
EliteCAD	11 Sp1
Facility Online	3.51
Ifc Engine Viewer	
IfcObjCounter	2.91
IfcStoreyView	2.1
IfcViewer	2
IfcWalkThrough	2.1
IfcXMLKonverter	1
MagiCAD	
Revit Build 2008	2008 SP1
SCIA-ESAPT	2001
Solibri Model Checker	4.2
Solibri Model Viewer	
Tekla Structures Precast Concrete Detailing	13
Tekla Structures Standard Design	13
Tekla Structures Steel Detailing	13
Vectorworks	2008 SP2
Vizelia IFC-VRML Viewer	

Source: National Institute of Building Sciences (NIBS)

software applications support open-standard data formats such as the Industry Foundation Classes (IFCs) of buildingSMART International (see Chapter 3), but many licensees don't know it, or have never attempted to use the IFC data format to exchange project data, and even fewer have used it as a standard data storage medium. Table 2.1 shows a list of applications that have been certified as compliant with IFC release 2x3 as of September 2008. More are continually being certified. An open certification process, in and of itself, results in greater insight and knowledge about interoperability. Shortcomings were identified in the initial process, which will be corrected in support of an improved process for future releases. For an up-to-date list of IFC-compliant applications, go to www.iai.hm.edu/ifc-compliant-software.

Even if a firm is unable to use IFCs to exchange information for its immediate business processes, routinely archiving building information data files in both their native formats and in IFC format can help ensure accessibility of the data long after the original data files can no longer be accessed because of file format changes in the original software. Additionally, nearly all software applications allow users to save data files in one or more proprietary data formats other than the native file format. While some of the original data may be lost when these "Save As" and "Import/Export" features are used, an audit of the data exchange capabilities of a firm's existing software applications can reveal the extent to which existing software tools will support at least some degree of interoperability. In some cases, the "dumbing down" of the data that occurs with these types of data conversions may be turned to a firm's advantage by helping protect against undetectable alteration of the original data files. A firm also may discover that, in addition to increasing its options for data exchange with business partners, these capabilities can be exploited for internal information exchanges between a firm's own design or business software applications.

Data exchange capabilities, both open standard and proprietary already exist in many software applications. Making effective use of this technology—which software licensees have already paid for—requires little more than exploring and testing the capabilities of existing software.

A second element of responsible information stewardship is maintaining rigorous data creation, filing, and archiving protocols. Well documented processes are the key to making them work. If the procedures and protocols are not documented, they will occur haphazardly, if at all. Enhancing a firm's exchange capabilities could be as simple as developing consistent procedures for file naming, data storage, data indexing, and data archiving so that information can be easily retrieved and validated. Most data files have searchable metadata tags (the file "properties") that are very valuable for data management but are rarely used. (Metadata is discussed in greater detail in Chapter 6.)

Some firms are successful in enforcing file naming, filing, and archiving conventions through standard operating procedures, but it is an unmanageable problem for many firms, as it depends on consistent human behavior. Software tools designed to enforce data nomenclature rules have long been available. More recently, enterprise-wide information-management software applications such as Newforma Project Center have emerged with full-text indexing and search capabilities for all file types, including e-mail, CAD, and BIM files, providing far greater "intelligent" access to proprietary data than previously possible. As a prerequisite to BIM implementation or as a strategic business goal in its own right, improved information management is likely to generate significant early returns in client service and employee productivity—without the significant investment in software and training for deploying new BIM technology.

The immediate business benefit of such simple steps is enhanced access to your organization's own information. You may discover that you have the ability to exchange information internally among different software applications that you did not previously know, or that information created for previous projects may now be exploited more effectively for future projects. Improved information management also may provide better protection against data loss and enhance business continuity. Finally, an increased awareness of the challenges of data exchange may enable you to articulate your data exchange needs more clearly to your software providers. The role of such feedback in the advancement of technology should not be underestimated.

A third step in implementing a systems approach to information management and inculcating an attitude of information stewardship is to initiate a dialogue within your organization about business processes, data sets, data formats, data validation, and electronic information exchange. Workflow and information flow within organizations are often horribly inefficient, inconsistent, and more difficult to change than external business processes, because they often involve requiring key people in the organization to change long-standing patterns of inefficient behavior. Your organization's data exchange capabilities with external business partners will improve dramatically if you begin by identifying opportunities for improved *internal* workflow and information exchanges, whether electronic or otherwise. Focus on eliminating redundant or repetitive processes. Conduct pilot information exchanges to compare and validate electronic information against traditional information exchange methods. Identify information exchange and workflow gaps and develop strategies for closing them. The knowledge you gain by improving internal workflow and information exchange, in addition to making your organization more efficient, will be invaluable for improving external exchanges.

ASSESSING TEAM CAPABILITIES

An organizational assessment of information management capabilities can be accompanied, or followed by, a collaborative assessment of the information management and information exchange capabilities of business partners. Project teams regularly form and dissolve, so the degree to which a collaborative assessment can occur and new business processes implemented may vary. But even the briefest business relationships can benefit from a regular dialogue about streamlining business practices.

A good basis for this joint effort is determining which party is the best authoritative source for a particular piece of information, and what pieces of information each authoritative source needs to provide to others to enable those third parties to perform their tasks. Each team member should analyze what it does, what information it handles, and whether it is the optimum "responsible party" for that information.

This dialogue begins informally and gradually becomes more structured and intensive as the size and complexity of the information to be exchanged increase. The scope of the dialogue might include:

- New types of building information a team member may be able to share that might be useful to others
- Which types of information, if provided by others, could help a team member perform its functions better
- How information is used in each team member's business processes and how it flows through their business systems
- Opportunities for frequent, "intermediate" information exchanges that might reduce or eliminate the number of low-value data entry tasks performed by team members
- Opportunities to eliminate overlaps or redundancies
- Information exchanges that might accelerate iterative workflow cycles

In the planning and design stages of a building project, in particular, a great deal of information that might be useful by other team members for preliminary decision making, comparative analysis, or iterative design purposes is withheld due to liability concerns or fear that it will be inadvertently incorporated into the final design. The problem is one of agreement as to the nature, quality, and appropriate use of such preliminary exchanges, and can be negotiated. For information in this category, project teams can execute a series of carefully circumscribed pilot data exchanges. As team members develop confidence in their mutual understanding of the data, they can expand the scope of such "no fault" exchanges, backed up, if necessary, by mutual indemnification agreements.

MANAGING EXPECTATIONS

Any change in business practice must be accompanied by an equitable adjustment in risk, accountability, and compensation. For example, one of the potential benefits of exchanging BIM data between an architect and a contractor is a reduction in the time needed for quantity takeoffs. But an architect might be reluctant to share the building information model for this purpose, out of concern that a building information model might be sufficiently complete to convey design intent but not sufficiently detailed or complete for quantity takeoff purposes. At the end of the construction period, an as-built BIM model is of considerable potential value to the building owner for facility management, real asset management, and operations-related purposes. But the owner and constructor or design-build team might have a different understanding of the degree to which the "as built" model represents real-world conditions, and that gap in understanding may represent an unacceptable risk to the builder.

It is not hard to imagine scenarios in which legitimate differences in understanding could lead to disputes and litigation. All parties involved with a building can agree that any information created about that building is useful to others and should be conserved, but it is extremely important that the parties have an explicit understanding regarding the scope, completeness, precision, accuracy, and appropriate use of any information exchanged. When information exchanges take place at significant milestones in the building lifecycle—when the active role of an information author ends—provisions must be agreed upon to indemnify the authoring party for any losses that may occur as a result of the inappropriate use or inappropriate reliance on the information transferred, which is no longer within the author's responsible control.

It is perfectly reasonable for the parties to a "milestone exchange" to agree to certain qualitative and quantitative standards for the information to be transferred, provided that the authoring party is appropriately compensated for any additional effort required to meet those standards that exceeds the original information needs of the authoring party or the original agreement between the parties. The best time to reach this agreement is at the time of the original agreement between the parties, so that information can be most efficiently gathered, updated, and conserved during the original process for which it was created. Too often, this issue is addressed in agreements only in the most cursory manner, requiring the authoring party to convey "a building information model" with no explicit understanding as to what the content of the model will be. If the parties are unable to define the content of the model at the time of the original agreement, then the principle of "no fault" transfer of "found" information should prevail, and the receiving party should indemnify and hold harmless the authoring party for any use of the information beyond its originally intended purpose.

Authoring parties also may be concerned about suddenly assuming responsibility and liability for information that historically has been the responsibility of others. In our first example, an architect might understandably have concern that a contractor will rely on the model—and hold the architect accountable for—material quantities. This would increase the liability of the architect for information that has been, historically and appropriately, the responsibility of the contractor. The transfer of this information from the architect to the contractor—intended to make the contractor's job easier—should not result in increased liability to the architect.

The fundamental dilemma is one of information assurance, except that in this case, it is not merely about the integrity of the electronic data or the verification of the data against an objective standard. Rather, it is about the core realms of expertise of the two parties, and who is best qualified to create which information. The resolution of this dilemma points strongly in the direction of early collaboration of the design and construction team. The constructor needs to impart enough construction knowledge to the architect to enable the architect to prepare a building information model suitable for quantity takeoff purposes, or the parties need to agree to a handoff of "responsible control" of the model at some appropriate interval (and with appropriate indemnification) so that the constructor can add quantity takeoff information to the architect's design intent model. Innovative early adopters are testing both methods.

As with any innovation, pilot testing is an important component in developing mutual information assurance. For example, an architect and a contractor could agree to an initial takeoff exercise in which a quantity takeoff of the building information model is completed independently of a quantity takeoff completed by conventional methods. The purpose of the exercise is to help both the architect and contractor gain greater insight into how to modify their modeling and business processes so that they both have confidence that the type and quality of information generated from the model is suitable for quantity takeoff purposes. The overriding goal is to streamline business processes across organizational boundaries and to enhance the profitability of both organizations. The business arrangements between the parties might subsequently be modified to compensate the architect for any additional effort required to create a richer information model, to enable the contractor to participate in the development of the model, or for both parties to share in the costs and benefits of jointly developing the model. Any change in business practice would be accompanied by appropriate indemnifications to ensure that neither party assumes any new, uncompensated, or inappropriate risks.

MEASURING PROGRESS TOWARD STRATEGIC GOALS

There is always an element of the unknown to the deployment of new technologies. Metrics can be difficult to establish for the deployment of a technology such as BIM that involves business relationships, enterprise workflow, project delivery methods, staff skill and training, and the design process. It is still possible, however, to establish goals and define objective metrics for measuring progress in BIM implementation. Not all goals and metrics can be expressed in dollars and cents, but they almost always can be quantified in some way that can be tied, at least indirectly, to the bottom line. The stronger the connection between an organization's BIM implementation strategy and profitability, the better the results of the BIM implementation are likely to be.

The Capability Maturity Model (CMM) of the National Building Information Modeling Standard (NBIMS)[1] is a good first step toward establishing BIM implementation benchmarks (see Table 2.2). The NBIMS CMM[2] is designed to measure the "maturity" of a building information model and the processes used to create it.

The use of the word *model* is an unfortunate choice of term here, adding yet another shade of meaning to a word that is already overused in this context. The term is borrowed from the software industry. It was originally developed in 1986 by the Carnegie Mellon Software Engineering Institute (SEI)[3], a federally funded research and development center, as a compendium of principles and practices for assessing the ability of government contractors to perform a contracted software project. The CCM concept has since been applied to related disciplines and activities such as software engineering, system engineering, project management, software maintenance, risk management, system acquisition, information technology (IT), and personnel management, and through NBIMS is now being applied to building information modeling.

To minimize confusion, the NBIMS CMM would be more aptly named the *Capability Maturity Index,* since that is what it truly is: an *index,* or benchmark, for measuring the *maturity* of your organization's BIM *capabilities.* It identifies eleven categories of maturity, each of which can be scored on a scale

Table 2.2 Capability Maturity Model.

Maturity Level	A Data Richness	B Life Cycle Views	C Roles or Disciplines	G Change Management (CM)	D Business process (BP)	F Timeliness/Response	E Delivery Method	H Graphical Information	I Spatial Capability	J Information Accuracy	K Interoperability/IFC Support
1	Basic Core Data	No Complete Project Phase	No Single Role Fully Supported	No CM Capability	Separate Processes Not Integrated	Most Response Info manually re-collected—Slow	Single Point Access No Information Assurance (IA)	Primarily Text[nd]No Technical Graphics	Not Spatially Located	No Ground Truth	No Interoperability
2	Expanded Data Set	Planning & Design	Only One Role Supported	Aware of CM	Few Business Processes Collect Info	Most Response Info manually re-collected	Single Point Access w/Limited IA	2-D Non-Intelligent As Designed	Basic Spatial Location	Initial Ground Truth	Forced Interoperability
3	Enhanced Data Set	Add Construction/Supply	Two Roles Partially Supported	Aware of CM and Root Cause Analysis (RCA)	Some Bus Process Collect Info	Data Calls Not in BIM But Most Other Data Is	Network Access w/Basic IA	National CAD Standard (NCS) 2-D Non-Intelligent As Designed	Spatially Located	Limited Ground Truth—Int Spaces	Limited Interoperability
4	Data Plus Some Information	Includes Construction/Supply	Two Roles Fully Supported	Aware CM, RCA and Feedback	Most Bus Processes Collect Info	Limited Response Info Available In BIM	Network Access w/Full IA	NCS 2-D Intelligent as Designed	Located w/ Limited Info Sharing	Full Ground Truth—Int Spaces	Limited Info Transfers between Commercial Off-the-Shelf (COTS) Software
5	Data Plus Expanded Information	Includes Constr/Supply & Fabrication	Partial Plan, Design, & Constr Supported	Implementing CM	All Business Process Collect Info	Most Response Info Available In BIM	Limited Web Enabled Services	NCS 2-D Intelligent As-Builts	Spatially located w/Metadata	Limited Ground Truth—Int & Ext	Most Info Transfers between COTS
6	Data w/Limited Authoritative Information	Add Limited Operations & Warranty	Plan, Design, & Construction Supported	Initial CM process implemented	Few BP Collect & Maintain Info	All Response Info Available in BIM	Full Web Enabled Services	NCS 2-D Intelligent And Current	Spatially located w/Full Info Share	Full Ground Truth—Int And Ext	Full Info Transfers between COTS
7	Data w/ Mostly Authoritative Information	Includes Operations & Warranty	Partial Ops & Sustainment Supported	CM process in place and early implementation of RCA	Some BP Collect & Maintain Info	All Response Info From BIM & Timely	Full Web Enabled Services w/IA	3-D—Intelligent Graphics	Part of a limited GIS	Limited Comp Areas & Ground Truth	Limited Info Uses IFC's For Interoperability
8	Completely Authoritative Information	Add Financial	Operations & Sustainment Supported	CM and RCA capability implemented and being used	All BP Collect & Maintain Info	Limited Real-Time Access From BIM	Web Enabled Services—Secure	3-D—Current and Intelligent	Part of a more complete GIS	Full Computed Areas & Ground Truth	Expanded Info Uses IFC's for Interoperability
9	Limited Knowledge Management	Full Facility Life Cycle Collection	All Facility Life-Cycle Roles Supported	Business processes are sustained by CM using RCA and Feedback loops	Some BP Collect & Main In Real Time	Full Real Time Access From BIM	Netcentric Service Oriented Architecture (SOA) Based w/Common Access Card (CAC) Access	4-D—Add Time	Integrated into a complete GIS	Comp GT w/Limited Metrics	Most Info Uses IFC's for Interoperability
10	Full Knowledge Management	Supports External Efforts	Internal and External Roles Supported	Business processes are routinely sustained by CM, RCA and Feedback loops	All BP Collect & Maint in Real Time	Real Time Access w/ Live Feeds	Netcentric SOA Role Based CAC	nD—Time & Cost	Integrated into GIS w/Full Info Flow	Computed Ground Truth w/Full Metrics	All Info Uses IFC's for Interoperability

Source: National Institute of Building Sciences (NIBS)

of one to ten. Version 1 of NBIMS acknowledges that the scale values of the CMM are subjective and in need of further definition and refinement, but the eleven categories appear to address all of the relevant information management and development categories of a building information model. The scale values are useful even in their initial draft state of development, particularly if an organization defines the values more precisely for its own purposes.

In late 2007, the NBIMS Testing Team, led by Professor Tammy McCuen of the University of Oklahoma and Air Force Major Patrick Suermann, P.E., Rinker Scholar at the University of Florida, conducted a test of the NBIMS CMM by evaluating the BIM maturity of the 2007 American Institute of Architects (AIA) Technology in Architectural Practice (TAP) BIM Award winners. An important part of the test was to measure the variance in scores between individual evaluators independently scoring each project. The degree of variance would be an indicator of how consistently the CMM rating scale could be applied to the same project by different evaluators, and therefore, a measure of how useful the CMM could be to the industry as an objective measure of BIM maturity. Though refinements were made to the NBIMS CMM as a result of the exercise, the variance in scores did not exceed 5 percent in any instance, and frequently varied by no more than 1 or 2 percent.

The eleven NBIMS CMM categories and their summary descriptions are as follows.

Data Richness. Refers to the degree to which a building information model encompasses the available information about a building. The scale ranges from individual pieces of unrelated data to information that is sufficiently comprehensive and authoritative to be regarded as corporate knowledge (see Table 2.3).

Life Cycle Views. Refers to the degree to which a building information model can be viewed (and used) appropriately by any players throughout the building life cycle who may have need of the data to execute their responsibilities (see Table 2.4). The current scale presumes that building data originates in the planning and design phase of a building life cycle, and measures the number of available views cumulatively from early planning stages through facility management/operations, then beyond "building specific" professionals to real estate portfolio managers, business operations managers, and external users such as emergency first responders. The greater the number of life cycle views supported by a building information model, the less likely that building information will be redundantly entered into separate information-management or business-process systems. This category of maturity has enormous implications for building owners, because it measures the degree to which building information can be transformed into business information.

Table 2.3 Data Richness Capability Maturity Model, Detail View.

Maturity Level	Data Richness
1	Choose this selection when you have established a BIM, but have only very basic data to load.
2	As you become more advanced, additional data will be available and be entered. This is still early in the maturity.
3	At this point you are beginning to rely on the model for basic data.
4	This is the first stage when data is turned into information.
5	The data is beginning to be accepted as authoritative and the primary source.
6	Some metadata is stored and information is typically best available.
7	Most users rely on information as reliable and authoritative; little additional data checking is required.
8	The information has metadata and is the authoritative source.
9	Limited Knowledge Management implies that KM strategies are in place and authoritative information is beginning to be linked.
10	Full Knowledge Management implies a robust data-rich environment, with virtually all authoritative information loaded and linked together.

Source: National Institute of Building Sciences (NIBS)

Table 2.4 Life Cycle Views Capability Maturity Model, Detail View.

Maturity Level	Life Cycle Views
1	Data is gathered as it is available but no single phase is authoritative or complete.
2	Since basic initial data is collected during planning and design, this is typically the first phase to be made available, but can be any phase such as construction.
3	An additional phase is available, typically construction; however, the two phases do not necessarily need to be linked.
4	A third phase is added; although information does not have to be flowing, it is assumed that some is.
5	A forth phase of the facility life cycle is added and some information is flowing.
6	An additional phase is added and clearly information is flowing to operations from the design and construction phases.
7	Information collected during earlier phases is flowing to operations and sustainment.
8	A cost model is supported and costs are linked to the information related to al phases. Life cycle costing can be performed.
9	All phases of the life cycle are supported and information is flowing between phases.
10	External information is linked into the model and analysis can be performed on the entire ecosystem of the facility throughout its life.

Source: National Institute of Building Sciences (NIBS)

Table 2.5 Roles or Disciplines Capability Maturity Model, Detail View.

Maturity Level	Roles or Disciplines
1	Roles apply to people's jobs, and at this level no one's role is fully supported through the BIM.
2	Roles apply to people's jobs, and at this level there is one person's role that is fully supported through the BIM.
3	Roles apply to people's jobs, and at this level there are at least two people's roles that are partially supported through the BIM but they still have to go to other products to accomplish their jobs.
4	Roles apply to people's jobs and at this level there are at least two people's roles that are fully supported through the BIM in that they do not have to go to other products to accomplish their jobs.
5	People's jobs in planning and design are fully supported through the BIM in that they do not have to go to other products to accomplish their jobs.
6	People's jobs in planning, design, and construction are fully supported through the BIM in that they do not have to go to other products to accomplish their jobs.
7	People's jobs in planning, design, construction are fully supported and operations and sustainment are partially supported through the BIM in that they do not have to go to other products to accomplish their jobs.
8	People's jobs in planning, design, construction, and operations and sustainment are fully supported through the BIM in that they do not have to go to other products to accomplish their jobs.
9	All facility-related jobs throughout the life cycle of the facility rely solely on the BIM to accomplish their jobs.
10	All facility-related jobs both internal and external to the organization rely solely on the BIM to accomplish their jobs.

Source: National Institute of Building Sciences (NIBS)

Roles or Disciplines. Refers to the number of building-related roles or disciplines that are accommodated in the modeling environment, and thus is a measure of how well information can flow from one role or discipline to another (see Table 2.5). The scale recognizes that currently available modeling environments are unable to accommodate even one role or discipline fully. The lowest end of the scale is partial accommodation of a single discipline, rising incrementally up to an environment in which all building-related disciplines can rely on the building information model as the sole information resource to perform their jobs. Like the Life Cycle Views scale, this scale presumes that building data originates in the planning and design phase of a building life cycle, and measures the number of roles/disciplines supported cumulatively from early planning stages through facility management/operations.

Business Process. Refers to the degree to which business processes are designed and implemented to capture information routinely in the building

Table 2.6 Business Process Capability Maturity Model, Detail View.

Maturity Level	Business process
1	Business processes are not defined and therefore not used to store information in the BIM.
2	Few business processes are designed to collect information to maintain the BIM in the organization.
3	Some business processes are designed to collect information to maintain the BIM in the organization.
4	Most business processes are designed to collect information to maintain the BIM in the organization.
5	All business processes are designed to collect information as they are performed.
6	All business processes are designed to collect information as they are performed but few are capable of maintaining information in the BIM.
7	All business processes are designed to collect information as they are performed and some are capable of maintaining information in the BIM.
8	All business processes are designed to collect information as they are performed and all are capable of maintaining information in the BIM.
9	All business processes are designed to collect and some maintain data in real time.
10	All business processes are designed to collect and maintain data in real time.

Source: National Institute of Building Sciences (NIBS)

information model as an integral part of each business process (see Table 2.6). This is a key, long-term metric of progress in BIM implementation, one that should be a strategic focus of every BIM software company. Whenever information can be gathered as an integral part of a business process, the compilation of that information is achieved at no additional cost. Whenever data is compiled as a separate process, the cost is greater and resources are diverted from primary business processes, reducing the likelihood that the data compilation task will be completed consistently. Or to put it another way, any time you have to take time out of your day job to compile data for someone else to use, the chances that you will do it consistently, if at all, are slim. The scale ranges from "business processes undefined and not used to compile data" to "all business processes are designed to collect and maintain information in real time." The high end of the range is a very high standard of performance to achieve.

Change Management. Refers to the degree to which an organization has developed a documented methodology for changing its business processes (see Table 2.7). Whenever a business process is found to be flawed or in need of improvement, a formal, documented process is followed that begins with a "root cause analysis" followed by a modification of the business process based on the analysis. The scale ranges from "no evidence of documented change management" to an environment in which business processes are routinely

Table 2.7   Change Management Capability Maturity Model, Detail View.

Maturity Level	Change Management
1	No change management process awareness is evident, nor has it been implemented in the organization.
2	There is an early awareness of the need for business process definition and change management in the organization, although implementation is not yet initiated.
3	Early implementation of business process definition is underway, there is an early awareness of the need for business process definition, and there is an awareness of change management and the need for root cause analysis in the organization.
4	Business processes are in place and there is an understanding of the full change management requirement to include root cause analysis and implementation of a feedback loop.
5	Business processes are in place and the organization has begun implementing change management procedures.
6	Business processes are in place and early change management processes are identifying changes, but no process is in place to make changes.
7	Early implementation of change management is in place and some processes are being maintained through a root cause analysis process.
8	Implementation of a change management process is in place and is beginning to be exercised, but is not fully endorsed by all participants.
9	The change management processes are in place, but are not efficient, and changes typically take more than 48 hours.
10	A mature and fully operational change management process is in place and process changes are implemented within 48 hours.

Source: National Institute of Building Sciences (NIBS)

supported by an integrated change management process that includes root cause analysis and feedback loops to assess the effectiveness of the change.

Delivery Method. Refers to the robustness of the IT environment to support data exchange and information assurance (see Table 2.8). The scale ranges from "BIM is only accessible from a single workstation with no integral information assurance" to "BIM is in a netcentric Web environment, delivered as a service in a service-oriented architecture (SOA), with role-based Common Access Card (CAC) enabled to enter and access information."

Timeliness/Response. Measures the degree to which BIM information is sufficiently complete, up-to-date, and accessible to users throughout the life cycle (see Table 2.9). The scale ranges from "information is collected when needed to respond to a question" to "information is continually updated from live-feed sensors and accurately reflects real-world conditions; responses to questions are immediate and authoritative."

Table 2.8 Delivery Method Capability Maturity Model, Detail View.

Maturity Level	Delivery Method
1	The BIM is only accessible from a single workstation and has no information assurance built in.
2	The BIM is not on a network but there is control over who can access the BIM.
3	The BIM is on a network and there is basic password control over data entry and retrieval.
4	The BIM is on a network and there is control over data entry and retrieval.
5	The BIM is in a limited Web environment typically found in a single office environment; IA is not in place to control data entry or retrieval.
6	The BIM is Web enabled but IA is not in place, although there is some control to access of the information. This environment would be found in a single office/company.
7	The BIM is in a Web environment so multiple people can operate on it and there is role-based IA manually controlled.
8	The BIM is in a Web-enabled environment and is considered secure. It is not in an SOA.
9	The BIM is in a netcentric Web environment and is served up as a service in a service-oriented architecture and CAC enabled but roles must be managed manually.
10	The BIM is in a netcentric Web environment and is served up as a service in a service-oriented architecture with role-based CAC enabled to enter and access information.

Source: National Institute of Building Sciences (NIBS)

Graphical Information. Refers to the degree of sophistication or embodied intelligence of graphical information (see Table 2.10). The scale ranges from "no graphics in the BIM; text only" to "graphical information stored in the BIM is object-based, parametrically intelligent, and includes information related to time and cost."

Spatial Capability. Refers to the degree to which the building information model is spatially located in the real world according to Geographic Information Systems (GIS) standards (see Table 2.11). This metric has implications for users across the building life cycle, including energy design and analysis, authoritative coordination with public infrastructure such as water and other utilities, and timely response by emergency first responders. The scale ranges from "the facility is not spatially located" to "information from the BIM is fully recognized in the GIS environment, including support for full metadata interaction."

Information Accuracy. Measures the degree to which information reflects real-world conditions (see Table 2.12). The scale ranges from "no ground truth; information is loaded manually, not verified electronically" to "all spaces are

Table 2.9   Timeliness/Response Capability Maturity Model, Detail View.

Maturity Level	Timeliness/Response
1	Information is re-collected when needed to respond to a question—the process is slow and un-automated and has to be reinvented each time a question is asked.
2	Most of the information needed to respond to a question must be collected to respond to the question; however, there is awareness of how to obtain the information.
3	Most information is in the BIM; however, many responses to data calls involve collection of data, which is then stored in the BIM.
4	Information is stored in the BIM and many data calls can be answered with information that is already in the BIM.
5	A significant portion of the response information related to a facility is stored in the BIM.
6	Responses to data calls related to the facility are primarily stored in the BIM.
7	All emergency response information is in the BIM and that is considered the primary source of accurate information.
8	Information stored in a BIM is available real time and although not from a live feed. Processes are in place to maintain its accuracy.
9	The information is stored in a BIM and is current enough to be a reliable source for information in an emergency.
10	Information is continually updated and available from live feeds to sensors. Responses to questions are almost immediate and are accurate and relational.

Source: National Institute of Building Sciences (NIBS)

Table 2.10   Graphical Information Capability Maturity Model, Detail View.

Maturity Level	Graphical Information
1	There are no graphics in the BIM, only text.
2	2-D drawings are stored in the BIM but there is no interaction with information; the drawings were not developed with the NCS.
3	The drawings stored were developed with NCS yet are still nonintelligent and not object oriented.
4	The drawings are 2-D but are intelligent—a wall recognizes itself as a wall with properties but they are as designed and not as built.
5	The drawings are 2-D and are intelligent—a wall recognizes itself as a wall with properties and they are as built but not current.
6	The drawings are 2-D and are intelligent—a wall recognizes itself as a wall with properties and they are current.
7	The drawings are 3-D object based and have intelligence.
8	The drawings are 3-D object based and have a process in place to keep them current.
9	Time phasing has been added to the drawings so that one can see historical elements as well as being able to project into the future.
10	The drawings stored in the BIM are intelligent and object-based and include time and cost information.

Source: National Institute of Building Sciences (NIBS)

Table 2.11 Spatial Capability Maturity Model, Detail View.

Maturity Level	Spatial Capability
1	The facility is not spatially located using GPS or GIS.
2	A basic location has been established using GPS so that one can locate the facility spatially.
3	The facility is recognized in a worldview spatially but no information is shared between the BIM and GIS.
4	The facility is spatially located and some information is shared with the GIS environment.
5	The facility is spatially located and information can be shared with the GIS environment although it is not integrated and interoperable.
6	The facility is located spatially and there is full information sharing between the BIM and GIS.
7	The BIM has been partially integrated into the GIS environment.
8	Information from the BIM is recognized on a limited basis by the GIS.
9	Information from the BIM is partially recognized by the GIS environment and some metadata is available.
10	Information from the BIM is fully recognized by the GIS environment, including full metadata interaction.

Source: National Institute of Building Sciences (NIBS)

Table 2.12 Information Accuracy Capability Maturity Model, Detail View.

Maturity Level	Information Accuracy
1	There is no ground truth and information is simply loaded into the system manually or unverified electronically.
2	There is some electronic validation of information for internal spaces.
3	Space is calculated electronically and not stored as a separate data element for internal spaces.
4	Internal spaces are identified electronically and some outside information is electronically calculated.
5	Many spaces and items are identified electronically yet some items are still entered manually, both internally and externally.
6	All internal and external spaces are identified electronically.
7	Internal spaces are computed electronically and some outside information is electronically calculated.
8	All units are calculated electronically and reported. If a polygon changes shape, then the updated information flows throughout the model.
9	All internal and external areas are computed and some metrics have been established to track compliance.
10	All spaces are calculated automatically and metrics are used to ensure information is available and accurate.

Source: National Institute of Building Sciences (NIBS)

Table 2.13   Interoperability/IFC Support Capability Maturity Model, Detail View.

Maturity Level	Interoperability/IFC Support
1	There is no interoperability between software programs. Information is reloaded for each application.
2	There is some interoperability but it is not automatic or seamless. Information may be cut-and-paste at this level of maturity.
3	There is some machine-to-machine flow of information but it is not common or the norm; it is still the exception.
4	Information is flowing between COTS products, often by using products from the same vendor. The interfaces are likely proprietary.
5	In this level of maturity, information is transferred between COTS products typically from the same vendor, but not all applications are supported.
6	There are good machine-to-machine linkages at this level of maturity and information interoperability is the norm.
7	Industry Foundation Classes are used on a limited basis for interoperability with some software packages.
8	IFC use is becoming more commonplace yet is still less often used than other approaches.
9	IFC use is the norm, but not exclusively used to attain interoperability. One would expect about 70–90% IFC-based interoperability.
10	At this level of maturity, IFCs are fully implemented and used for interoperability.

Source: National Institute of Building Sciences (NIBS)

calculated automatically and methodologies are in place to ensure that information is accurate." This is a significant factor in determining the level of confidence one has in the information. If we lack confidence in the information, we are destined to re-collect it repeatedly during each phase of the building life cycle.

Interoperability/IFC Support. Measures the degree to which data can be reliably exchanged among software applications using the open-standard Industry Foundation Classes (see Table 2.13). The scale ranges from "no interoperability between software applications" to "IFCs are fully supported and used for information exchange." While any interoperability approach may work on a small scale, the only currently viable, open international standard is IFC.

TOWARD A NEW BUSINESS PARADIGM

What we have described thus far in this chapter is the essence of *information modeling.* It is about designing a reliable *system* for compiling and exchanging information in a culture of *information stewardship.* It has little to do, per se, with individual software applications or even technology. It is a mindset—a way of thinking. It is about recognizing that one's workflow and work product is part of a continuous workflow, a *business system*. It is often asserted that the building industry is too fragmented, too hidebound by traditional business practices, too adversarial, and too litigious for any tangible improvements in efficiency and productivity to occur. Too often, however, these concerns are raised in defense of inaction. The problem, it is said, is so big, so complex, and so overwhelming that no single individual or organization, regardless of its size, can do anything about it. "Someone else has to fix it" is a common refrain that can be heard throughout the building industry.

The building industry is indeed fragmented, hidebound, adversarial, and litigious. These characteristics may be endemic, but they are not immutable. Nor do they present significant or even challenging obstacles. It is ironic that while few of us think of our business processes and work products as part of a system, many of us regard our existing business culture as a system so deeply entrenched that we are powerless to change it. The problems in the building industry are of our own making. As Albert Einstein once said, "We cannot solve problems by using the same kind of thinking we used when we created them." Ultimately, leaders will emerge who will capitalize on the benefits and lay the foundation for substantive change in our industry.

ENDNOTES

1. National Institute of Building Sciences, "National Building Information Modeling Standard, Version 1, Part 1: Overview, Principles, and Methodologies," http://www.wbdg.org/bim/nbims.php, ed.

2. National BIM Standard Project Committee, "National BIM Standard Capability Maturity Model Workbook," National Institute of Building Sciences, http://www.wbdg. org/bim/nbims.php

3. "Capability Maturity Model," Wikipedia, http://en.wikipedia.org/wiki/Capability_Maturity_Model.

CHAPTER 3

Business Process Reform

It is not necessary to change. Survival is not mandatory.

—William Edwards Deming

Implementation of building information modeling and integrated project delivery present something of a chicken-and-egg dilemma for every business leader in the building industry. The full benefits of BIM will be realized only after most of the industry has made the transition from current technology and business practices, because the design, construction, operation, and ongoing maintenance of every building is accomplished by a broad network of people and organizations who need to work together to exchange information in a coordinated fashion. Until then, business leaders and managers have to determine to what extent they can efficiently and profitably integrate BIM into their own operations.

There is an inherent conflict between the "all for one, one for all" collaborative spirit of the new business paradigm and the need of individual businesses and organizations to continue doing business, earn a profit, and differentiate themselves from their competition. Proponents of BIM and IPD tend to trumpet the former and overlook or neglect the latter. A more realistic approach recognizes that collaboration and competition can coexist, and that innovation depends on the dynamic tension between the two.

Competition propels individual firms to develop best practices. When widely shared and adopted, best practices become standard operating procedure and the entire industry operates at a new, more advanced level. At the moment, individual firms are competing to develop best practices in integrated project delivery, and by doing so are differentiating themselves from firms that remain mired in the old ways of doing business. In recent years, a critical mass of consensus has developed that business practices need to change, which helps lower the risk for these pioneering firms. But the level of risk that pioneers are prepared to assume may be too high for many companies. Every organization needs to determine the level of "innovation risk" it is prepared to assume. Every company needs to find the right balance and chart its own course of business process reform and technological development in the context of overall industry trends and its own market position.

MANAGING INNOVATION RISK

Successfully managing change and the risks of change are every business leader's greatest challenges. Organizations at the forefront of change do not necessarily assume any greater liability risks, but they most certainly assume greater financial risks, because the industry as a whole may not be mature enough for their early investments in new technology and new business processes to pay off for a very long time. Conversely, those who wait too long may find themselves at a significant competitive disadvantage, unable to meet the new needs and demands of the marketplace. Figure 3.1 depicts a hypothetical relationship between innovation and individual business enterprises in the building industry. The vertical scale represents the number of businesses in the industry, while the horizontal scale represents their relative degree of innovation. A small number of firms, the pioneers, are at the leading edge of innovation. These

FIGURE 3.1
Building Industry
Innovation Scale

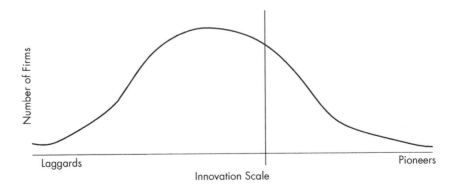

companies explore many avenues of innovation with the understanding that only some will yield direct, measurable, and tangible results. Their culture of innovation, however, has a "branding benefit" that distinguishes them from their competition in the marketplace and helps attract the best talent. These firms have honed "first mover advantage" to a sharp edge and know how to monetize innovation—or convert concepts into cash—quickly enough to offset the risks of being pioneers.

An equally small number of firms—the laggards—are well behind the pack. These notoriously risk-averse companies adopt innovations in business practices and technology only when the innovations have become mainstream. Their strategy, whether they articulate it or not, is to pay for innovation at commodity prices. Sticking to the path of the tried-and-true may seem like a safe strategy, but it carries the significant hidden risk of falling too far behind the market and being unable to catch up quickly enough. What may seem like the safest course of action may be the riskiest of all.

The bulk of the industry is somewhere in the middle, following directly behind the pioneers like a surging wave. The bold vertical line on the graph, about two-thirds of the way across the innovation scale, is the optimum position for most organizations: just behind the pioneers and early adopters—where one can take advantage of the lessons the pioneers have learned—and just ahead of the great wave of the vast majority. At this point on the innovation scale, the high risks of experimental practices and technologies are mitigated, while substantial opportunities still exist to differentiate oneself from the competition.

This recommendation is not unlike the advice you might receive from a good financial advisor. The conventional wisdom of personal financial planning is that young investors should tolerate a higher level of risk to benefit from a higher level of capital growth over their lifetime, while older investors should reduce their level of risk to ensure a secure and steady income in retirement. As a business leader, you want to stake out a position of moderate to moderately high innovation and risk tolerance. Why? Because innovation is as vital to business growth and development as calculated financial risk is to wealth and prosperity. Staying just ahead of the crowd allows you to leverage promising innovations to your competitive advantage while safely sidestepping technological developments that may not yield results.

THE IMPERATIVE OF CHANGE

There is an additional reason why the safest strategy can turn out to be the riskiest: in business as in biology, the status quo is never a viable option. Capitalism, like life itself, is sustained by a continuous cycle of birth, growth, development, and decay. At any given moment in time, your business is either on an upward trajectory or a downward trajectory. Every organization must adapt to change merely to sustain itself. Like a swimmer treading water in the middle of a lake, any sense of equilibrium derived from the status quo is temporary and illusory. The only viable options for the swimmer are to swim to shore or to drown. The only viable options for businesses are to adapt to change or die.

The world around you is in a continuous state of change. Even if you have no desire to grow your business, you need to be responsive to change just to survive. The economy is cyclical. Markets expand and contract. Products become obsolete. Technology advances. Costs go up. New competitors continuously enter your markets. Client organizations merge or dissolve. Employees come and go. Client contacts change jobs or retire. Simply put, some aspect of the business environment in which you operate today will not be the same tomorrow. A culture of innovation enables an organization to function not as a single organism, doomed to die, but as a complete ecosystem, capable of adapting to changing conditions and continuously regenerating itself. Healthy biological ecosystems—those most able to adapt to change—often appear most stable. Companies can develop a similar capacity for successfully managing change, which we will discuss later in this chapter.

INNOVATION MANAGEMENT STRATEGIES

So how can you keep your organization in the "sweet spot" of innovation with respect to BIM implementation, just behind the pioneers and ahead of the crowd on the innovation/risk scale? The idea that you can "manage" innovation may seem counterintuitive, but a few simple strategies can keep an organization at the cutting edge and away from the bleeding edge:

- Continuously evaluate emerging technologies and business practices. There is always something new on the horizon, and by keeping tabs on what's happening you will quickly develop the critical framework that will enable you to distinguish truly useful tools and methods from technological dead ends.
- Continuously seek out opportunities to operate more efficiently, both internally and with external project team members.
- Identify those technologies (including but not limited to BIM) and practices that will strengthen your core competencies, broaden your service

offerings, streamline workflow, increase the quality of your work products or services, lower costs, improve team communication, and enhance electronic information exchange.

- Deploy only those technologies that satisfy one or more of the above criteria.
- Establish metrics at the time of deployment to measure progress toward your goals. If you find it too difficult to measure a return on investment in dollars and cents, establish other metrics that measure the financial return indirectly, such as a higher hit rate on proposals or reduced cycle time for completing certain operational or project tasks.
- Deploy only mature, commercially available technologies that require only minimal customization (if any) to be used productively in your organization. Beware of software that requires extensive customization. When you have to pay for customization before you can make use of the product, you are effectively underwriting the cost of software development for an under-capitalized company.
- Deploy technologies six to twelve months after their initial introduction to the marketplace. The time delay allows bugs in new releases to be worked out by others and, more importantly, allows the technical support knowledge base of the technology provider to mature, so that any problems that do arise can be resolved promptly.
- Avoid experimental or noncommercial technologies, unless they have a demonstrated track record of success and are backed by an organizational infrastructure capable of supporting their continued development.
- Participate in beta testing programs only if you have a serious interest in helping to shape the development of the software *and* have sufficient technical expertise to make a meaningful contribution.

This innovation management strategy will require financial investment, education, and training, and a "shake out" implementation or transition period during which productivity may temporarily drop. But it will keep financial risks to a minimum and enable you to exploit emerging technologies and business practices to maximum advantage. Be aware that the "early adopter" competitive advantage of any innovation strategy is temporary. As the industry catches up, the individual tools and practices you've deployed will give you less and less of an edge over the competition. But the culture of innovation you will have fostered in your company and the market perception of your organization as innovative and forward thinking—both far more significant competitive advantages—will be permanent.

THE "I" IN BIM

Implementation of building information modeling may not be as difficult or challenging as it first seems. Most building information—and therefore most of the data contained in a building information model—is alphanumeric; it consists of words and numbers. This is true even in the design and construction stage of buildings, when the geometrical and spatial representation of building information is most important. Because we too often think of BIM in terms of geometry, we tend to lose sight of the "I" in BIM, which is information. Kimon Onuma, FAIA, president of Onuma, Inc., and creator of the Onuma Planning System, frequently makes the point that an Excel spreadsheet of spatial data is a building information model.

Whenever information is organized into something as simple as a table, with column headings that define the type of content contained in each cell in that column, the information can be described as "structured," and can be manipulated "intelligently" by simple algorithms designed to qualitatively distinguish the values in one column from the values in any other.

Many constructors rely heavily on Excel for construction cost estimating, despite the availability of sophisticated construction cost estimating software. Perhaps not coincidentally, respondents to the FMI/Construction Management Association of America (CMAA) Eighth Annual Survey of Owners[1] rated Microsoft Excel as their most important software application. If so many industry professionals across the building life cycle rely so heavily on spreadsheets for compiling data critical to their work, then perhaps we need to change our perception of what a building information model is, and focus more attention on efficiently and reliably exchanging a type of data that is right at our fingertips.

Alphanumeric data is much easier to compile and organize than geometrical or spatial data, and much of it is sitting around waiting to be compiled and organized. Consider that for almost any building designed and constructed in the past ten years, virtually all relevant information about that building—most of it alphanumeric—was originally created in digital form by someone, whether the author was a product or equipment manufacturer, an architect, an engineer, a construction specifier, a construction cost estimator, a contractor, a subcontractor, or a fabricator. At some point—typically when the information is conveyed from the authoring organization to another party—most of that digital information is reduced (degraded) to nondigital form. This is done for reasons of expediency: to facilitate existing, paper-based business transactions and practices, or to get around contractual, business, or electronic barriers to electronic information exchange. At least conceptually, all of that information is available for compilation into a coherent digital database. Significant barriers to its efficient use and exchange remain, but such a database—if it could be compiled and validated at a reasonable cost—would have considerable value. For this reason, firms throughout the industry should constantly evaluate opportunities to compile, conserve, and exploit such information—pulling the needles out of the haystack—whenever possible.

BUSINESS REFORM STRATEGIES

By far the most important yet least addressed aspect of implementing BIM is the corresponding change of business practices needed to optimize the opportunity afforded by BIM, whether the practice in question, such as integrated project delivery, requires the cooperation of business partners or is entirely internal to your firm. The aspects of business process reform that should be considered as part of any BIM implementation effort include:

- Greater electronic information exchange to reduce or eliminate manual data entry whenever possible
- Reduction of cycle time whenever and wherever possible to reduce or eliminate low-value or no-value tasks such as processing Requests for Information, preparing shop drawings, or measuring existing conditions
- Reduction of time spent on "defensive documentation" primarily intended to protect the author, and reallocation of that time toward value-added tasks that advance the interests of the project
- Integration of construction cost estimating with building information modeling to generate more precise, accurate, and detailed estimates earlier in the project delivery process
- Deployment of building information modeling technology and expertise for constructability analysis and construction sequence planning
- Reorganization of business processes to enable more tasks to occur concurrently rather than sequentially
- Increased prefabrication of construction assemblies
- Implementation of direct design-to-fabrication processes
- Automated, real-time monitoring and analysis of operating systems and equipment to achieve and maintain optimal performance
- Greater comparative analysis of the predicted performance of systems and equipment with their actual performance, in order to improve the quality of future design analysis forecasts.

These changes are only partly about technology. BIM and related technologies may enable these reforms to occur, but the fundamental issue is how we use information to improve the way we do business. Our fragmented business processes do not allow information to flow freely. Everything about our existing business model—contracts, license obligations, and customary business practices—evolved with the express purpose of compartmentalizing information to make it manageable using available paper-based information-exchange methods. We are no longer restricted by those methods and so should not restrict ourselves to the business practices developed to accommodate them.

Information authorship—who creates which information—is another aspect of our business process that business leaders need to consider seriously. Data re-entry (often multiple times) has become so routine that we no longer think about it. Whenever information is entered into a software application, it is worth asking whether the information has already been gathered by someone else for some other purpose, or whether it could be created more efficiently and accurately by someone else. Information is likely to be most accurate the first time it is gathered and stored by the person or organization that most urgently needs it. Information authorship should migrate toward its most logical authoritative source, regardless of traditional or conventional practice. The goal is to optimize the entire business process for a given project or facility, not to optimize the internal processes of a single organization. Every company should develop the habit of routinely analyzing its information flow to determine when information should be created internally and when it should be retrieved from others. As part of that process, systematic procedures should be implemented for retrieving and validating information received from others. In later chapters, we will address issues of information assurance in greater depth, including the importance of metadata (which is nothing more than data about data; think of a card catalog in a library) as a first element of data validation.

INDUSTRY-WIDE REFORM EFFORTS

As we indicated at the beginning of this chapter, there are two components of building industry business process reform: industry-wide changes in business practices, and changes in each individual organization to adapt to the new business model. The "chicken" in our chicken-and-egg dilemma is beyond the realm of control of any individual company or organization. These are aspects of innovation that the industry must address collectively by establishing open standards or by developing and disseminating best practices that eventually become routine operating procedures or practices.

The need for collaboration and collective effort to spur innovation is undeniable. As we noted in Chapter 1, the best examples or applicable models are not to be found in manufacturing but in economic sectors that more closely match the profile of the building industry. Consider, for example, barcode labeling and scanning technology in the consumer product economy, an economic sector that is far more "fragmented" than the building industry. First developed to identify and track freight railroad cars in the 1950s, a modified form of the original barcodes—known as the Universal Product Code (UPC)—was later developed for the retail grocery industry, and still later extended to all consumer products.[2] For such a system to be implemented successfully, millions of product manufacturers, wholesalers, and retailers had to work together. It required the complete overhaul of inventory management systems, product labeling and pricing methods, and installation of new point-of-sale equipment. The success of the system depends entirely on the integration and interoperability of these formerly separate business processes and technologies. It also depends on ubiquity. It has to be possible to label any product offered for sale with a UPC; otherwise, the anomalous items requiring manual processing would seriously erode the efficiency of the entire system. Those of us old enough to remember when the technology was first introduced can recall some of the bottlenecks to UPC ubiquity: items purchased in bulk from wholesalers and sold to retail customers in smaller units, such as board feet of lumber or meat by the pound. But even these obstacles were quickly overcome, and we now accept that anything sold anywhere by anyone can have a UPC label slapped on it.

As we noted in Chapter 2, the disparity in productivity growth between the building and agricultural industries can be attributed, in part, to the disparity in government-funded statistical data collection and research. That distinction is absent when the building industry is compared to the consumer product industry. The system of Universal Product Codes dramatically increased productivity, efficiency, and the quality of inventory information throughout the consumer products supply chain without government assistance. The initial standards-development effort was led by a coalition of six grocery industry associations of product manufacturers and retailers, and the cost of its development and implementation was borne entirely by the industry itself.[3]

The story of the development and implementation of the UPC system should dispel any notion that the building industry is not able to meet the challenges to greater efficiency, productivity, and interoperability that it faces. The building industry is far less fragmented, involves far fewer players, and has a much shorter end-to-end supply chain than the consumer economy. On the other hand, the information supply chain challenges we face are greater than product labeling and inventory control. The data sets we need to exchange are more complex, and our supply chains and information flows are neither linear nor durable. Project teams continuously form and dissolve; the links in our supply chains are continuously reforged, unlike the business relationships between wholesalers and retailers, which are maintained for far longer periods. The flow of information in the building industry also tends to occur independently of, or parallel to, physical workflow activities or tasks, and cannot always be tied to physical objects in quite the same way as a UPC label can be attached to a product. Nevertheless, the implementation of the UPC system in the consumer products industry is an excellent model for interoperability and industry-wide cooperation in the building industry. At the very least, implementation of UPC or Radio Frequency Identification (RFID) tagging technology could do much to enhance the quality, completeness, and currency of building information models.

INDUSTRY STANDARDS AND INNOVATION

Standards-development efforts present every industry with something of a paradox. Standards can foster innovation by providing a common environment for further technological or business environment, or, if not developed carefully, can inhibit innovation or constrict the business environment. Whether a standard is prescriptive or performance-based may be a factor in its effectiveness, but even that is not always clear. The UPC system is highly prescriptive, even rigid. Individual players in the consumer product industry must adhere to the UPC format if they wish to take advantage of its benefits; they cannot invent their own barcode system. But the UPC system does not inhibit individual players from driving further innovation in their own supply chains. Wal-Mart exploited UPC and other inventory control technologies in ways that other retailers did not, by leveraging the vast database of information generated by UPC labeling to its own advantage. Wal-Mart analyzed the voluminous data that their inventory system generated, developed detailed institutional knowledge of regional and seasonal buying patterns, and used that knowledge to automate and drive the efficiency of their inventory control, purchasing, and product pricing systems. UPC technology was available to every company in the consumer product supply chain, but only one very small company figured out how to exploit the technology to become one of the largest and most profitable retailers in the world.

Barcode technology is an excellent example of a highly prescriptive technology standard that benefits an entire industry while providing an open platform for further innovation to those players with the vision to perceive greater opportunities.

The Leadership in Energy and Environmental Design (LEED) Green Building Rating System™ of the U.S. Green Building Council is a voluntary, consensus-based standard to support and certify successful green building design, construction, and operations.[4] LEED is also highly prescriptive, assigning points to individual characteristics of a building according to their environmental impact or benefit. As the science and technology of green building advances and experience with the rating system increases, however, the prescriptive nature of the LEED system becomes problematic. Buildings cannot receive points for environmentally friendly characteristics that may not have been anticipated by the rating system. Professionals debate whether the system gives the proper weight to the environmental benefits of things such as the historic preservation or adaptive reuse of existing buildings. And finally, the rating system cannot be applied uniformly to buildings of all types. As a result, competing rating systems or standards are beginning to proliferate, and the USGBC itself has developed different LEED rating systems for different building types.

These two examples illustrate the formidable challenges of standards development. The UPC system requires every player in the consumer products market to adhere to a rigid format for product identification and labeling, but significantly improves productivity throughout the industry while allowing the most enterprising players to excel. It is durable yet flexible, requiring little or no revision since it was first introduced. It requires virtually no interaction with the standards-development body by the individual organizations that use it. The LEED system, a type of checklist-based certification system, requires periodic revision as green building technology develops. It requires intensive interaction with the standards-setting body for certification, which can become a bottleneck as the number of buildings for which certification is sought increases. It creates little or no opportunity for individual players to distinguish themselves in the marketplace it tends to inhibit, rather than spur, innovation.

The comparative analysis of these two standards is imperfect. They serve vastly different purposes and address very different problems. Sustainability in building design, construction, and operations cannot be achieved with something as simple as a barcode. But the lesson for building industry business leaders is that standards-development efforts are important. There is value in paying attention to—and considerable risk in ignoring—technology-related standards-development efforts. Building information modeling and integrated project delivery are in their very early stages of maturity. The opportunity to influence standards-development efforts will never be greater. There are significant business opportunities for companies like Wal-Mart that recognize the value of emerging technology standards and figure out how to leverage them to achieve greater success in the marketplace.

THE INDUSTRY STANDARDS LANDSCAPE

The scope and diversity of building industry standards-development efforts may seem bewildering at first. Multiple initiatives by numerous organizations are underway, which may seem confusing to the casual observer. Among the reasons for so many simultaneous efforts is that there are many challenges that need to be addressed, so individual industry groups with a particular area of expertise or professional interest set out to identify and resolve particular problems. While it's helpful to divide the work into manageable pieces, there is a risk that problems will be addressed in a piecemeal fashion at the expense of comprehensive systems solutions. In the absence of an academic research and research-publishing tradition through which knowledge can be disseminated quickly and widely, multiple research and standards-development efforts could result in duplication of effort or significant knowledge gaps.

The way out of this dilemma is for individual problems to be addressed in an industry-wide context, for individual efforts to be conducted as openly as possible, and for individual groups to coordinate with one another as closely as possible. In recognition of this, the North American chapter of the International Alliance for Interoperability (IAI), a council of the National Institute of Building Sciences (NIBS), reorganized itself as the buildingSMART alliance, with a broad mission to serve as a clearinghouse and forum for coordination of the work of many different research and standards-development groups. The buildingSMART alliance now coordinates a growing number of U.S.-based research and development projects that are managed, led, and funded by a dozen different public and private organizations. Similar reorganizations are underway in other IAI (now buildingSMART International) chapters worldwide. A complete list of buildingSMART alliance programs and projects is available on the alliance Web site, www.buildingsmartalliance.org, but it is worth highlighting a few of them here.

An early project of buildingSMART International is the open-standard data format known by the collective moniker of Industry Foundation Classes, or IFCs. The IFC data format is designed to facilitate interoperability—the seamless, reliable exchange of building information between software applications. Support for the IFC format has been incorporated into dozens of software applications worldwide, including most of the popular BIM applications in use in the United States today, and has had a significant impact in raising awareness of the importance of interoperability, whether through the IFC format or other means.

Interoperability is a foundational technology for greater efficiency and productivity in the building industry, just as the UPC system is the foundation for greater efficiency and productivity in the consumer product industry. But while interoperability is very important, it is only one of the many challenges facing the industry.

The National BIM Standard (NBIMS), one of the primary buildingSMART alliance initiatives, aims to build upon the foundation of IFC interoperability to improve the planning, design, construction, operation, and maintenance of buildings by fostering the use of a standardized, machine-readable, and comprehensive building information model in a format that is useable and accessible to all players throughout the life cycle of the facility.[5] Another building SMART effort, the agcXML Project, managed by NIBS and funded and led by the Associated General Contractors of America (AGC), will result in a set of extensible markup language (XML) schemas to allow the digital exchange of the transactional construction data now commonly exchanged in paper-based contract documents and forms. Some, though not all, of the data in the agcXML schemas is building information that could be incorporated into a building information model. By structuring this transactional data for the first time, agcXML provides a mechanism for capturing this data efficiently.

In a related effort, FIATECH, an industry consortium of owners, engineering construction contractors, and technology suppliers, seeks to fast-track the development and deployment of technologies—BIM or otherwise—to improve how capital projects and facilities are designed, engineered, constructed, and maintained. FIATECH defines the "capital projects industry" as the economic sector that executes the planning, engineering, procurement, construction, and operation of predominantly large-scale buildings, industrial plants, facilities, and infrastructure. Founded in 2000 as a separately funded initiative of the Construction Industry Institute (CII) at the University of Texas at Austin, FIATECH in late 2004 released the ambitious Capital Projects Technology Roadmap (CPTR). It lays out a comprehensive strategy and detailed tactical plans for achieving the vision of "fully integrated and highly automated project processes with radically advanced technologies across all phases and functions of the project/facility life cycle."[6] While the member organizations of FIATECH have specific needs related to the scale of their projects, much of the development work envisioned by the Roadmap will be applicable to the building industry at large. The CPTR is arguably the first detailed, comprehensive statement of the technical and standards-development work that needs to be done to streamline business processes throughout the building industry.

ALIGNING BUSINESS STRATEGIES WITH INDUSTRY STANDARDS

The FIATECH Roadmap identifies nine critical components of the virtual enterprise of the future and includes detailed tactical plans to advance technology and business process reform for each component:

- Scenario-based project planning
- Automated design
- Integrated, automated procurement and supply network
- Intelligent and automated construction job sites
- Intelligent, self-maintaining, and self-repairing operational facilities
- Real-time project and facility management, coordination, and control
- New materials, methods, products, and equipment
- Technology- and knowledge-enabled workforce
- Life cycle data management and information integration.

The nine components of the Roadmap and their relationships are depicted in Figure 3.2. While the timelines for completing the tactical plans for each component extend as far as seven years (through 2011), the nine components and their relationships as depicted in Figure 3.2 are useful today to individual organizations for mapping out their own technology strategies within an overall industry context. As noted in the previous chapter, the Capability Maturity Model of the National Building Information Modeling Standard (NBIMS), even in its initial draft form, is a tool that the business leaders throughout the building industry can use to advance their building information modeling efforts. Similarly, companies can use the FIATECH Roadmap—or the product of other industry standards-development efforts—as a guide for aligning their proprietary strategies with industry-wide efforts. If a company knows, for example, that the FIATECH tactical plan for scenario-based project planning is approaching completion, that company can choose to leverage the knowledge created by the industry group rather than replicating the effort on its own. Conversely, if the FIATECH tactical plan for intelligent and automated construction sites is far from completion, a company could elect to join industry colleagues on the FIATECH tactical team to accelerate the effort or choose to

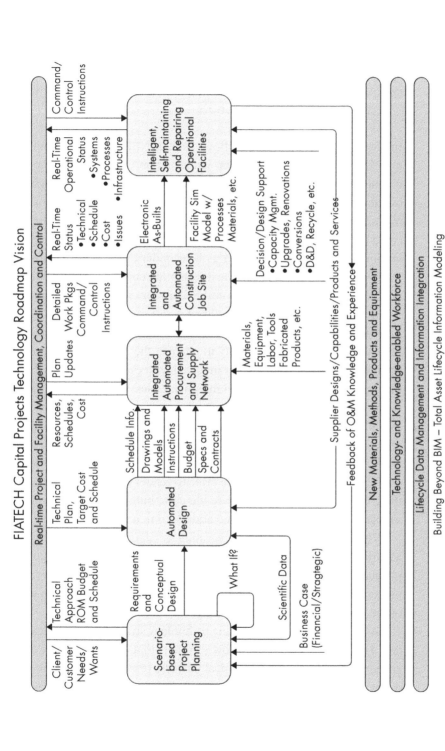

FIGURE 3.2 FIATECH Capital Projects Technology Roadmap Diagram. Source: FIATECH. Used by permission.

delay their own proprietary implementations in that area until greater industry standardization has been developed.

 ## INTEGRATING INFORMATION GATHERING INTO THE BUSINESS PROCESS

Implicit in industry standards efforts such as NBIMS and the FIATECH Roadmap is the idea that information gathering and preservation should be integrated with mission-critical business processes. Architects and engineers design buildings. Constructors build them. Owners and facility managers operate and maintain them. The focus of each player at each stage in the building's life cycle is the building itself. The priority is getting the immediate job done. The building information itself is viewed as a secondary concern, a means to an end. Everyone may be eager to use the information contained in a building information model, but there is far less enthusiasm for creating and maintaining one.

If data entry has to be completed as a separate, unrelated step, it is not likely to be completed routinely, and the building information model will soon be out of date. This is not likely to change; the industry will never reach the point where every team member involved with every building dutifully updates the model for the sake of model integrity. Unless it is integrated into the work itself, maintaining the model is simply extra work. It may have benefits for overall productivity, but it does not improve individual productivity. It will be difficult to persuade the people "in the trenches" that they should do this extra work for the sake of the greater good. It takes only one unwilling or uncooperative player in the building life cycle to render a BIM out of date. To achieve the goal of consistently up-to-date life cycle building information models, this simple fact must be accepted and addressed in a systemic way.

The solution is tighter integration of data collection and compilation activities—BIM creation and maintenance—with the real-world tasks that people need to do every day. But if data entry is an integral part of every business process, the integrity of the data becomes mission-critical and the maintenance of the model becomes self-sustaining.

As a part of a forward-looking technology strategy, business leaders should look for technology and workflow solutions that exhibit this type of tight integration of data with mission-critical business processes, just as the UPC system drove tighter integration of previously independent inventory management, product labeling and pricing, and point-of-sale systems in the consumer products industry. Prior to UPC, grocery stores manually affixed a price sticker (or, in the case of canned goods, an ink stamp) on every item in the store. Part-time grocery store clerks (one of the co-authors of this book among them) marked the price on every item just prior to placing it on the store shelf, using a price book as a reference guide. The task was completely independent of inventory control, the productivity of cashiers, or the speed at which customers could get through the checkout line. It had to be managed entirely separately. When the prices on individual items changed—as they frequently did—the stickers or ink stamps had to be removed from every item and replaced with new ones. Today, the prices of items are maintained in the inventory control system, which matches the price to the UPC code for the product. To communicate prices to customers, a single computer-generated display sticker is affixed to the front of the shelf where the item is displayed, and the manual repricing of hundreds or thousands of individual items is a thing of the past. The formerly independent task is now tightly integrated with inventory management. The system is a critical element of cashier productivity and, when integrated with similarly-encoded customer "discount" cards presented at checkout, can be used to collect detailed information about individual consumer preferences and buying patterns.

When this technology was first implemented, it was widely believed that it would never work and that the public would never accept this new technology, because critical product information—the price of the product—would be separated from the product itself. When faced with frequent price changes, grocery clerks would default to the "easy" thing and slap a new price sticker on the cans of soup. Nimble-fingered cashiers wouldn't trust the barcode scanners, and would continue to manually key in the numerical codes, Customers wouldn't trust that the price sticker at the front of the shelf would match the price they actually paid. There was a transitional period during which all of these dire perdictions came true, but it passed so quickly that few remember it. The acceptance of UPC technology today is overwhelming.

The same type of tight integration of "data" with "product" has to occur if building information modeling is ever to fulfill its promise. For example, a maintenance engineer who replaces a defective piece of equipment should not be able to close out the work order without updating the building information model to reflect the replacement.[7] This will only happen if the building information model is an integral part of the facility management and maintenance software that the maintenance engineer uses to do his job. Similar business cases for tighter integration of data with real-world business processes could be described for business processes throughout the building life cycle.

Business automation can be accelerated further by engaging business partners in a continuous dialogue to increase the amount of information exchanged electronically, even for seemingly inconsequential business processes. A packing slip, for example—the final piece of documentation that allows a recipient to validate that the product ordered is the product received—invariably contains vital building information about the product, such as the manufacturer's name and model number, a product description, and a serial number. It may be the only piece of documentation that confirms beyond a reasonable doubt the identity of the product installed in a building. Yet this information is rarely, if ever, systematically collected, and even more rarely conveyed or collected in electronic form. An enterprising recipient might scan the information from a paper document, but the integrity of the data would then depend on the quality of both the optical character recognition (OCR) software and the physical scan itself. A better option is for recipients of this type of information—the customers of the product vendors—to insist that it be delivered in electronic form with the product, either as an RFID tag or a machine-readable specialized barcode. This one bit of technology, alone, could bring the industry significantly closer to the goal of up-to-date life cycle building information models at a minimal cost. It may actually result in a net savings when the costs related to generating, printing, handling, and discarding the current paper trail is taken into account. While the information would be equally accessible to everyone, it also would provide a platform for further innovation for enterprising players, who, like Wal-Mart, figure out how to exploit the information to gain a market advantage.

The obstacles to electronic information exchange are often not as formidable as many people think. Consumers of products and services in the building industry often fail to exercise the power they have as consumers, simply accepting what is given to them rather than insisting on what they need. Consider the degree to which banking—an economic sector in which data integrity is vital—has evolved more and more toward electronic transactions. Electronic banking—paying bills, receiving statements, and, recently, receiving electronic copies of cancelled checks—has been possible for years. But even the most high-tech banking customer had no alternative but to write and mail paper checks to those individuals or vendors who could not, or would not, accept payments electronically. Soon, however, some enterprising banks provided an online tool for sending paper checks, with the bank taking over the functions of printing and mailing. Some banks charge their customers no fees—not even the cost of postage—for this service. For retail bank customers, this left just one aspect of banking that could not be automated—depositing paper checks received from others. No matter how much customers had automated their banking processes, if someone sent them a paper check, they had to take it to the bank. Once again, a few pioneering banks acted to remove this final obstacle to 100 percent electronic banking: they provided their customers with an option to deposit paper checks electronically. Their customers simply scan paper checks with their own scanner or use a special check scanner provided by the bank. Customers of these progressive institutions can conduct all of their bank transactions electronically, *regardless of how the rest of the world does business.* Similar opportunities for streamlined workflow and reliable electronic information exchange are available to customers of all types in the building industry, if only we would assertively—and collectively—seek them out. Operating in a BIM environment is possible for any organization, whether or not business partners or the rest of the industry jump aboard.

LEADERSHIP AND VISION

The business process reform needed to achieve greater productivity and efficiency in the building industry can occur only through visionary leadership. As a business leader, you must be able to articulate a clear vision of the future of your company—how it is going to change and how those changes will improve your business. Readers who are not the senior leaders of their companies but who wish to implement the concepts presented in this book must first obtain the full understanding and support of the leaders of their company. It cannot be done any other way. The entire organization must make a long-term commitment to the vision, and that commitment begins at the top.

The vision should be written clearly and concisely. It should be ambitious yet achievable. It should be supported by a detailed strategic plan with defined goals, objectives, tasks, and metrics for measuring progress toward the goals. The strategic plan should be integrated into the annual budget-planning process. Periodic progress reports—at least quarterly—should be made to the board of directors, senior firm managers, and the entire staff. Everyone needs to be apprised of and held accountable for progress so that momentum can be maintained. Progress toward achieving defined goals should be included in performance evaluations.

One of the co-authors of this book had the privilege of witnessing visionary leadership in action about ten years ago while working for a previous employer. The author's immediate supervisor had been assigned the task of developing a vision for business transformation by the leader of the organization. The supervisor enlisted his team, including the author, in developing the vision, thus ensuring that his team members could contribute to, understand, and fully support the vision. The vision was presented to the leader of the organization, who endorsed it and instructed the development team to present the vision to the senior managers of the organization—all Ph.D's and renowned experts in their chosen fields—the following week. The leader of the organization introduced the team leader, who presented the vision and strategic plan. It was clear that some members of the organization were less than enthusiastic about it. Immediately following the presentation, the leader of the organization stood and informed the staff that this was the vision and strategic plan that the organization would follow for the next five years, and that he would personally assist anyone not interested in working to achieve the vision in finding a new job. The comment was greeted with a bit of shock, but there was no misunderstanding the depth of the leader's commitment. The vision and strategic plan were then communicated to the entire organization of nearly four thousand people via a glossy publication that enabled everyone to understand their role. The senior managers of the organization were apprised of progress every quarter. The vision consisted of twelve components, and each quarterly leadership briefing reported on the progress of each component. The presentation highlighted the starting point for each component, the end state that was to be achieved, and the progress achieved toward each goal to date.

The implementation did not always go smoothly. There was a bit of push back. Some people kicked, screamed, and whined the whole time. But the vision was achieved in three years instead of the originally scheduled five, and the vision development team—now the implementation team—documented a benefit/cost ratio of 3.3:1. As Figure 3.3 shows, the implementation team also forecast future returns of 2.9:1 to 5.6:1, depending on the action of future leaders. To determine the benefit/cost ratio, the performance of the organization was benchmarked before the implementation began. Operational costs were calculated, which were then apportioned according to the twelve components of the implementation strategy. To minimize bias, an independent organization tracked performance.

The footnote to this story is that the organization in question was a government organization, which added an additional dimension to the challenges of business process reform. If business can be transformed in a government organization with a professional union of highly talented and powerful people, then it can occur anywhere with effective leadership.

FIGURE 3.3
Benefit/Cost Ratio
Projection

Figure of Merit	Best Case	Most Likely	Worst Case
TBO (Total Benefit of Ownership)*	$90.9	$54.1	$47.5
TCO (Total Cost of Ownership)*	$16.5	$16.8	$17.9
NPV (Net Present Value)**	$70.0	$36.2	$28.0
B/C (Benefit Cost Ratio)**	5.6	3.3	2.9

ENGAGING BUSINESS PARTNERS

While the preceding example was not specific to implementing a BIM environment, it illustrates the type of leadership, strategies, and steps needed to achieve substantial business process reform. The vision of a BIM implementation is greater productivity, efficiency, and quality in all business processes throughout the life cycle of a building facility. The strategic goal for achieving the vision is greater information sharing. The objectives of a strategic implementation plan are twofold: deploying new technologies for sharing information and developing new business processes to ensure that information, once created, can be reused and repurposed for the remainder of the useful life of the building. The second objective requires that you engage in a dialogue with business partners, to ensure that information supplied to your company by external sources is provided in a usable format and that information you create is in a format that will be useful to others.

The dialogue with business partners and the implementation of digital information workflows across organizational boundaries may initially be challenging, but it is first and foremost a business problem and only secondarily a technical problem. Whenever information sharing is identified as a business imperative, business partners find a way to do it. When technical obstacles are perceived as insurmountable, they are rarely surmounted.

Manufacturing industries faced a similar challenge prior to implementation of "just in time delivery" of parts and supplies, which is now the norm in manufacturing, most notably in the automotive industry. Assembly lines had reached an optimum level of productivity and cost-effectiveness. Manufacturers began looking beyond the assembly line for ways to reduce costs, and the supply chain became a prime target. Maintaining large inventories of parts added no value to the final product. The costs of storage space, "shrinkage" due to damage or theft, and labor costs for the multiple handling of the same items were simply a drain on the bottom line. Manufacturers had traditionally maintained large inventories of parts and supplies because if a critical part were missing, an entire production line might have to shut down until the part was received. The manufacturers assumed full responsibility for having the right parts available at the right time. "Just in time delivery" transferred this particular responsibility for keeping production lines running to vendors and suppliers. To be sure, logistical and communication systems had to developed and implemented so that vendors had the information and tools they needed to meet their new obligations for timely delivery. But resolving these technical considerations was an outcome of the business decision, not the cause. The critical factor in implementation of the new system was renegotiating the terms and conditions of the business relationship with vendors and suppliers. Timely delivery became an integral part of the value proposition of the relationship, as much as the quality and quantity of parts and supplies.

Companies in the building industry who are recipients of building information from consultants and vendors are in the same position to negotiate the scope of their deliverables and to specify that information of a certain type and quality be included in the deliverable. As in manufacturing, the first step is to renegotiate the terms and conditions of the business relationship. Then new logistical and communication workflow systems must be established, including the scope and format of information to be exchanged. Vendors and consultants who are to be held accountable for delivering information must have the means to do so. They cannot be held accountable for providing information if the requirements are unclear or if it is beyond their ability to provide it due to circumstances outside of their control. The value proposition must be redefined to include appropriate compensation for any additional effort required. These are the elements of business process reengineering.

BUSINESS PROCESS MODELING

There have been many approaches to business process reengineering in the building industry in the past, and frankly, many did not work all that well. In all likelihood, the failures were probably not due to flawed processes so much as flawed implementations. Business leaders in the building industry, particularly in the design professions, are often reluctant to lead with "my way or the highway" determination. These organizations have a culture of collegiality and consensus that their leaders are often reluctant to disturb. Important business decisions are often communicated as strong suggestions, with no consequences for failure to implement them. To be sure, an extreme top-down leadership approach is as prone to failure as the consensus model. No business leader can reform an organization without the support of his or her people, support that is best earned through persuasion rather than coercion. An effective business leader, however, knows how to navigate a middle course between bottom-up consensus decision making and top-down, heavy-handed, single-minded direction. In far too many companies in the building industry, though, the pendulum has swung too far toward the consensus approach.

A second possible reason for the failure of prior business process reform methods is that business leaders may have recognized the need for change but did not understand how to transform their organizations. This is a particularly difficult challenge, as many of our business processes have evolved through tradition, a tradition that does not include managing change. A frightful number of business processes and workflows in the building industry are undocumented, even though "everyone knows how it's done." As an industry, we are simply not accustomed to thinking about and analyzing our business processes in a methodical way.

With building information modeling, we are collectively attempting—not entirely consciously—to use three-dimensional visualization to remedy our business process modeling deficiencies. The 3D visualization attributes of BIM enable project teams to achieve a much greater mutual understanding of an intended result. 3D representation of buildings also allows building design professionals to communicate more effectively with those team members—clients, especially—who are untrained in the esoteric and often impenetrable 2D language of plans, elevations, sections, and details. It is estimated that 90 percent of people who are presented with 2D drawings are unable to visualize the intended, physical result—a building or other man-made structure—without assistance, if at all.

The implications of this communication breakdown are enormous. Consider, for example, that most commercial buildings are required by law to display plan diagrams of emergency egress paths at conspicuous locations throughout the building. The vast majority of the occupants of buildings are unable to identify their office location on the diagram and how it relates to the nearest paths of emergency egress. The diagrams—intended to enhance life safety—are, in effect, useless. We can overcome these significant communication obstacles with the help of 3-D visualization, but this is only a first step; 3-D visualization alone will not transform the industry.

The ability for everyone involved in a building to achieve a much greater common understanding of the information they share appears to be fostering much greater dialogue about workflow. Suddenly, everyone seems to have an opinion about how things could work better. In a way, we are backing into the science of process engineering and the discipline of business process modeling. But rather than reinvent the wheel (again), we should look to and build upon the work done by others in these fields.

Business process modeling can be done many ways, and there is extensive documentation available about the various approaches (see Figure 3.4). Integrated Definition (IDEF)[8] modeling and Business Process Modeling Notation (BPMN)[9] are two common methods used for BIM-related business process modeling. In the building industry, IDEF appears to be giving way to the more popular BPMN. To date, buildingSMART International, the buildingSMART alliance, and the U.S. National BIM Standard Committee have standardized on BPMN. Whichever method you choose, many how-to books and Web-based

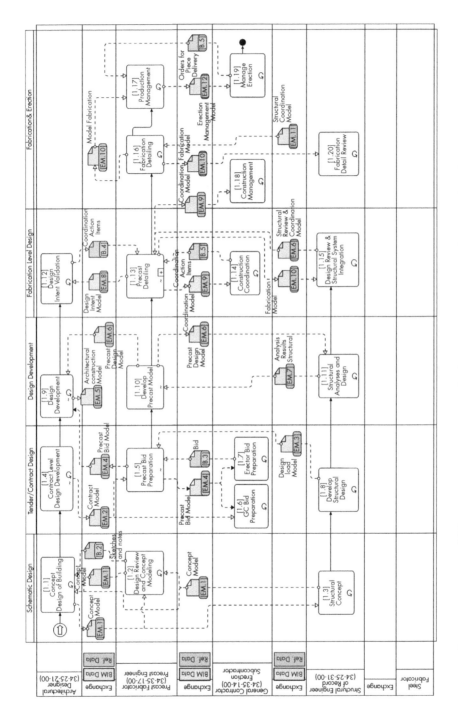

FIGURE 3.4 Draft Business Process Model for Precast Concrete Design and Fabrication. (Source: National Institute of Building Sciences [NIBS.])

information resources are available to learn how to use the modeling processes, and relatively little time is needed to become proficient in using them. Many software applications are available for graphically modeling business processes. Microsoft Visio Professional, for example, will support all of these business process modeling methods. Sophisticated business enterprise modeling, however, may require more robust software tools.

There is no limit to the level of detail to which business process modeling can be applied, but it is wise to set limits so as not to become ensnared in a web of inconsequential detail. While not a prerequisite to business process modeling, it is best to understand first how an existing business enterprise functions, which can be very revealing. This is known as modeling the "as-is" process. It is possible to model down to any level of detail within an organization, but detailed modeling is best done on an "as needed" basis for specific parts of the business enterprise; it is not necessary to model the entire organization to the same level of detail.

Business process modeling is not much different than putting together a construction schedule. Each organization has defined work activities with defined durations and a critical path for its workflow. Business leaders who are accustomed to using construction scheduling software may find it very easy to use business process modeling software. These programs conveniently allow detailed business processes to be rolled up into higher-level business summaries. Whichever tools are used, it is important to track the time and cost of the building process modeling effort, so that eventually the return on investment can be measured.

Once the "as-is" model is developed, the next step is to create the "to be" model. This is a little more challenging, as it requires the organization's leaders to imagine how their organization might function in the future. By developing the "as-is" model first, however, this exercise is a little easier. A profound understanding of the organization as it exists generates considerable insight about what it could be. This approach has a lot in common with the building design process, so it is likely to seem familiar to most building industry business leaders.

Once completed, the business process model becomes a roadmap that the organization can rely upon long into the future. In developing the process model, it is important to be both realistic about what is possible and ambitious about what can be accomplished. Implementation can be executed in phases, which allows business leaders to identify those intermediate implementation goals that are most likely to result in early benefits. These successes can then be successively built upon to implement fully the new business process model. Successful implementation will depend on the cooperation of many, if not all, of the people in the organization. Complex models may be no more useful in that regard than emergency exit diagrams. Care should be taken, as illustrated in the following example, to illustrate the process model both in a summary and detailed fashion, so that everyone understands the entire process at a conceptual level as well as their individual role or function at a detail level.

BUSINESS PROCESS MODELING CASE STUDY

Business process modeling was used in a recent buildingSMART alliance initiative to define the functional requirements of a Building Information Modeling standard for architectural precast concrete, by focusing on the multiple information exchanges between an architect and a precast contractor throughout the design, fabrication, and installation process.[10] The project was funded by the Charles Pankow Foundation and executed by a project team that included FIATECH, the architecture firm HKS Inc., High Concrete, Arkansas Precast, Georgia Institute of Technology, and Technion, the Israel Institute of Technology. The results are intended to be incorporated into the National Building Information Modeling Standard.

Figure 3.5 shows "Level 1" business process models for the "as-is" (current practice) and "to-be" (future practice) business processes for the design, fabrication, and installation of precast concrete. The project team also developed far more detailed workflow diagrams for each process, as shown in Figures 3.6 and 3.7.

Because this was a research study for which comparative results were desired, both processes were actually executed in parallel and the actual workflows recorded. Thus, the "to-be" process diagram is not conjectural, as it would be in a typical business process modeling exercise. The architectural design and engineering detailing for the precast concrete façade panels of a twenty-story commercial building were completed using conventional two-dimensional CAD tools. The same tasks were completed simultaneously and independently using advanced 3-D BIM tools. Each business process was studied in terms of workflow, information exchanges, and design productivity. The 2-D process required 440 hours for design and 830 for drafting, versus a total of 350 hours for 3-D modeling, for a net productivity gain of 58 percent.

The research project consisted of two additional parts, but this part of the experiment illustrates how business process modeling can be used to reform business processes in a methodical way. A few guidelines for developing business process models for integrated project teams include:

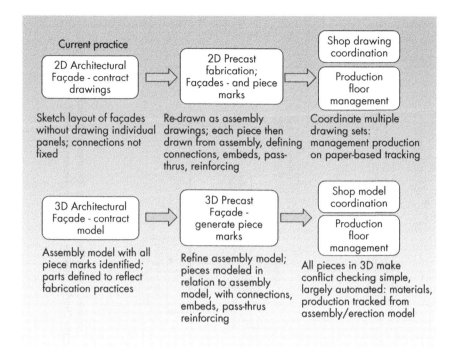

FIGURE 3.5
Business Process
Models, Architectural
Precast Concrete De-
sign and Fabrication.
(Source: National
Institute of Building
Sciences [NIBS.])

- A review of the business processes of others to identify similarities. No business process is unique. The closer your own business processes mirror industry practice, the more likely that commercially available software and other technology will be able to meet your business needs.

- Realignment of individual activities to occur in parallel instead of in sequence to dramatically shorten the cycle time for the overall business process.

- Organization of processes into modules or components. This helps support parallel processes and increases the number of processes that can be outsourced to third parties.

MANAGING CHANGE

Business process modeling is a vital tool of change management. Once the "as-is" process has been documented and the "to-be" process has been defined, the potential return on investment can be calculated by measuring the difference in operational cost between the two, then subtracting the cost of the business process modeling effort. The net savings, if any, is the value to the organization of the proposed change.

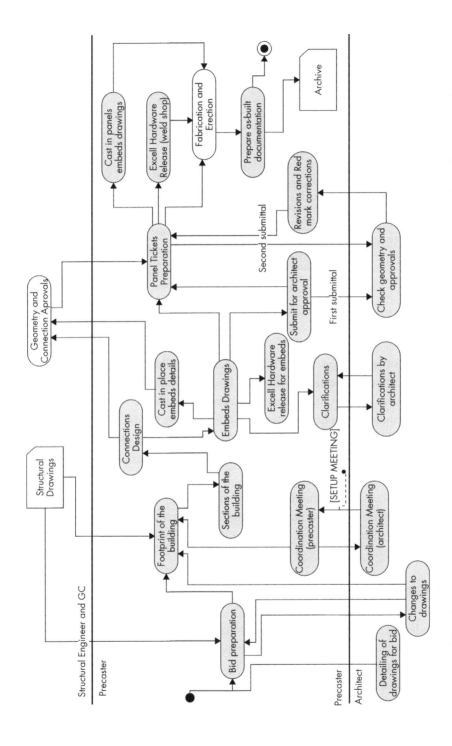

FIGURE 3.6 2-D Workflow for Architectural Precast Concrete. Source: R. Sacks, I. Kaner, C. Eastman, and D. Yang, *Building Information Modeling for Precast Concrete, Part A: Rosewood Experiment: Goals, Methods, Execution and Results.* Report to the Charles Pankow Foundation, November. 2007.

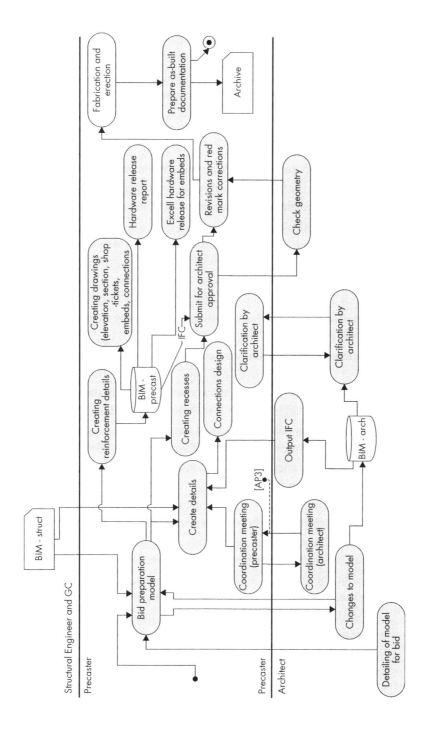

FIGURE 3.7 3-D BIM Workflow for Architectural Precast Concrete. Source: R. Sacks, I. Kaner, C. Eastman, and D. Yang, *Building Information Modeling for Precast Concrete,* **Part A: Rosewood Experiment: Goals, Methods, Execution and Results. Report to the Charles Pankow Foundation, November. 2007.**

The cost of continuing in the current mode of operation is the easier to determine, since that data should be readily available. Forecasting the operational cost of the new business model may be more difficult, but a good-faith estimate should be made, if for no other reason than to improve the organization's operational cost-estimating skill. As in the precast concrete example, it may be necessary to do some pilot testing in order to refine both the business model and the operational cost forecast. The goal is to manage change methodically by using a starting point—the "as-is" business process—as a benchmark for measuring results.

If the comparative analysis reveals a negative return, then the proposed business model is clearly flawed, and either the current process should be retained or another avenue should be pursued. The business process modeling exercise, however, has still been a success—the company has avoided pursuing an undesirable path and is more knowledgeable about its current operations and the range of available options. Flaws in a business process can also be addressed through another tool of change management: "root cause analysis." An in-depth analysis of specific problems can yield insights that lead to corrective action. Root cause analysis is also an opportunity to include others in the exercise as reviewers, who may identify even greater opportunities.

The process of making forecasts and comparing them to actual results is iterative and will improve each time, but only if the forecasts are made and documented. The odd thing about change is that in the midst of a change process we have a tendency to forget where we started, so we tend not to recognize progress. This is sometimes called the "microwave oven syndrome." A microwave oven might cook better and faster than anything we've previously used, but we wait impatiently as the seconds go by. We've forgotten how long it used to take to heat up a cup of coffee.

Measuring change brings the perception of value in line with actual value. When an organization fails to measure progress in any change process, resistance to change is likely to increase throughout the organization, even if the change is beneficial. We just don't remember how bad things were, so we don't see what the big deal is about where we are. The organization slowly slides back toward preserving the status quo. It reverts to an organism—doomed to die—rather than an ecosystem capable of adapting to changing conditions and renewing itself indefinitely. This phenomenon has played itself out in the building industry again and again. Yesterday's innovators become today's most intransigent resisters to new trends and technologies. Our goal with this chapter is to help today's resisters become tomorrow's innovators, and to help today's innovators avoid becoming tomorrow's resisters—in other words, to foster a culture of permanent change.

The utility and value of the business process model do not end with the completion of the initial transformation. No business process model is perfect, and, in any event, the world has continued to change since it was created. Ideally, the change process set in motion by the modeling exercise will encourage everyone in the organization to continue looking for ways to further improve the business process. How these new ideas are managed is extremely important. First, the original "to-be" model must become the new "as-is" model—the baseline for further change and improvement. Ideas then need to be thoughtfully incorporated into a new business process modeling exercise. This is the essence of change management: change happens regularly and methodically, not irregularly and haphazardly. When change is managed properly, "change fatigue" is eliminated; managed change becomes a regular part of doing business. Everyone in the organization understands what is changing, when it is changing, how they should do their own jobs, and how they should interface with others. Successive changes become easier to implement than the first, because the scope of change is typically less each time, and the organization is now acculturated to change. Failure to implement formal change management processes will quickly render most organizational change initiatives impotent. Change management introduces sustainability and helps ensure continuous improvement.

POPULATING THE BUILDING INFORMATION MODEL

When you define business processes, you also are documenting information flows. This is at the heart of the business process modeling effort. While business processes and information exchanges occur thousands of times each day in the building industry, they are rarely documented. Some, but not all, of the information embodied in these exchanges should be incorporated into the building information model. Part of the business process modeling exercise, then, should be to determine which information should be incorporated into the BIM, and which information is purely transactional. This brings the discussion of this chapter full circle: by documenting which of the required information exchanges in a business process should be incorporated into the BIM, the maintenance of the BIM is more likely to become an integral part of the business process.

ENDNOTES

1. FMI and Construction Management Association of America (2007), "Eighth Annual Survey of Owners." Retrieved July 29, 2008, from http://www.fmiresources.com/pdfs/07SOA.pdf.

2. Wikipedia (2008), "Barcode," retrieved August 5, 2008, from http://en.wikipedia.org/wiki/Barcode.

3. Ibid.

4. U.S. Green Building Council (2008), "LEED Frequently Asked Questions," retrieved August 11, 2008, from http://www.usgbc.org/DisplayPage.aspx?CMSPageID=1819#LEED.

5. National BIM Standard Project Committee (2007), *National Building Information Modeling Standard, Version 1, Part 1: Overview, Principles, and Methodologies* (Washington, DC: National Institute of Building Sciences).

6. Barton, S., ed. (2004), *Capital Projects Technology Roadmap.* Austin, Tex: FIATECH.

7. The integration of building maintenance and repair work orders with building information model maintenance was first suggested to us by George Korte, Vice President of Management Consulting, Total Resource Management, Inc.

8. Knowledge Based Systems Inc. (2006), "IDEF - Integrated Definition Models," retrieved August 7, 2008, from http://www.idef.com.

9. Object Management Group (2007), "Business Process Modeling Notation (BPMN) Information," retrieved August 7, 2008, from http://www.bpmn.org.

10. FIATECH (2007), "BIM for Precast Concrete." Retrieved August 7, 2008, from http://www.fiatech.org/projects/autodesign/bpc.htm#_projpartact.

CHAPTER 4

BIM-Based Enterprise Workflow

If you want something you've never had before, you're going to have to do something you've never done before.

—Zig Ziglar

In the previous chapter, we discussed business process (workflow) modeling. In this chapter, we look more closely at how some of these concepts can be implemented in your own business (see Figure 4.1). Building information modeling can potentially affect every aspect of a business enterprise. Therefore, BIM implementation is best viewed as an integral part of every business process, rather than as an isolated endeavor related only to a few specific tasks or projects. The fundamentals of business—productivity, cash flow, revenue, and profit—are the same for any for-profit business; only the details differ according to business type. The fundamentals of BIM implementation are also the same, though different types of businesses use building information for different purposes. The concepts outlined in this chapter apply to businesses of all types: architecture firms, engineering firms, architecture/engineering firms, general contractors, design-build firms, subcontractors, construction management firms, facility management firms, building owners, or even design-build-operate-and-maintain firms.

FIGURE 4.1
Enterprise Workflow:
How Does Your
Process Work?
(Source: Copyright ©
VISI/COINS. Used by
permission.)

BIM IMPLEMENTATION FUNDAMENTALS

BIM implementation will affect general business operations as well as the products or services your firm provides. Implementing these concepts uniformly—for both project and non-project operations—will help streamline business operations, establish a consistent working environment for your employees and business partners, and increase the percentage of your organization's total work effort that is devoted to value-added tasks. The following are some basic concepts of BIM implementation at an operational level.

Ensure that the data is entered only once during the building or information life cycle by the most authoritative source. This is the most significant step you can take to reduce non-value-added effort. Every time information is entered into a software application—in every facet of your business operations, not just projects—it is worth asking whether the information could have been obtained electronically from someone else who may have produced it with greater reliability and less effort. You may be surprised to discover the volume of information for which the authoritative source is someone or some company other than your own.

Even within organizations, the volume of information that is entered multiple times into different software applications, sometimes by the same people, can be staggering. Analyze the data routinely compiled in your existing software applications and eliminate as many of these overlaps as possible. Determine whether information can be exchanged electronically among these applications, or, if possible, designate one application as the sole repository of a particular type of data and eliminate the duplicative storage of that data entirely.

In relational database design, this is a formal concept known as *database normalization.*[1] There are three goals of normalization: minimize duplication of information, eliminate "logical inconsistencies," and safeguard against "data anomalies." In plain English, when data are stored in multiple locations, inconsistencies will develop over time, undermining the integrity of both data sources. Relational database software developers have developed sophisticated rules and procedures for database normalization. In the building industry, we can normalize a great deal of our data far more easily with simple business rules.

Send and receive data in the most structured electronic form possible. To the extent possible, encourage business partners from whom you routinely receive information to format their data so as to minimize the amount of pre-processing required to make the data usable in your organization. Take the same initiative to ensure that the information your firm generates and conveys to others is in a format that is most usable to the recipients with minimal degradation of content.

If two software applications are unable to exchange data directly, there may be an intermediate data format, either open or proprietary, that will make the exchange possible. Most software applications import and export data in multiple formats. Pilot testing—or a discussion with the software providers—may be necessary to determine the integrity and completeness of data sets transferred by any chosen method, but it is almost always worthwhile to take advantage of such digital data exchange capabilities.

Virtually all building information available today is originally created in a structured, digital form, including:

- Geometry
- Construction specifications
- Construction cost information
- Construction sequence and scheduling information
- Manufactured building products information
- Building systems information
- Building materials information
- Building performance information
- Fittings, furnishings, and equipment information

During the course of the life cycle of a building—and, especially, during building design and construction—much of this data is routinely degraded to nondigital form for only one reason: so that the information can be transferred from one party to another. Often, the sending party is entirely unaware of the digital capabilities of the receiving party. As a result, the information is degraded by the sender and reconstituted by the recipient for no reason at all. Take advantage of every opportunity to engage your business partners in a conversation about data exchange and to develop data exchange protocols at the most structured digital level possible.

Information authors may be unwilling to transmit digital information out of concern that they will be held liable for inappropriate or unintended use of the information or that they will assume liability for information contained in the data transmitted that heretofore has been the responsibility of the receiving party. The classic example is building material quantity takeoff information that easily can be extracted from a BIM model created by design professionals. Historically, accurate quantity takeoffs have been the responsibility of constructors, who would *infer* material quantities from printed construction documents prepared by architects and engineers that did not explicitly list material quantities. The constructor was solely responsible for the material quantities, unless the design professionals had failed to depict or otherwise document required equipment or materials properly.

This demarcation of responsibility has been rigidly maintained even for building components that are easily tallied, such as doors and windows. In the conventional business process, an architect prepares door and window schedules that list and specify each door and window by type, then labels each door and window on the plans and elevations with a symbol indicating the type. The construction cost estimator then laboriously counts each door and window on the drawings by hand. The audit and analysis capabilities of BIM are likely to embolden design professionals to assume responsibility for the quantities of building components that are expressed verbosely as objects in the model and can be tallied in simple arithmetical units at the object level, such as doors and windows, light fixtures and other wiring devices, plumbing fixtures and fittings, fire protection devices, and HVAC system components and controls. Essentially, anything that is purchased and installed as a unitary item can be expressed as an object in BIM and easily counted. Constructors, however, are likely to retain responsibility for materials measured by weight, length, area, or volume (and for which a waste factor must often be calculated), such as concrete, brick, tile, gypsum board, paint, architectural woodwork, ductwork, piping, finish floor materials, accessible ceilings, and membrane roofing. But even with only a partial shift of responsibility with respect to material and equipment quantities, design professionals and constructors should be able to agree, with appropriate indemnification, to allow the constructor to assume stewardship of the BIM model for quantity take-off purposes, for enriching the model with constructability information, or for construction sequence analysis.

Any existing business practices or risks that stand in the way of greater efficiency and productivity can be addressed through discussion, negotiation,

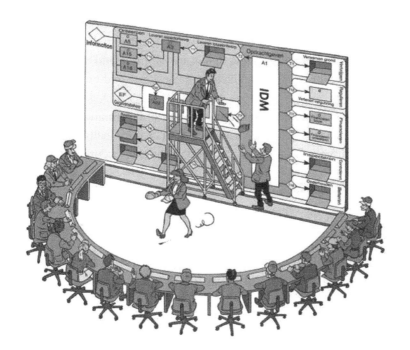

FIGURE 4.2
Developing a Business Process Model in a Collaborative Environment. (Source: Copyright © VISI/COINS. Used by permission.)

and agreement (see Figure 4.2). The parties to any project need to develop a mutual understanding of:

- Their mutual information exchange needs
- The nature of any information being transmitted
- The intended purpose for which the information is to be used
- The appropriate use of transmitted information by the receiving party

New standard forms of agreement recently published by the American Institute of Architects and ConsensusDOCS specifically address these issues and can be very helpful in clarifying the ambiguities in roles and responsibilities that have emerged as a result of these new data exchange capabilities.

In some instances, your firm may have no choice but to accept information in whatever format it is provided, including paper documents. While this is the least desirable situation, it is worth exploring electronic capture methods such as scanning and optical character recognition (OCR) for both text documents and drawings, and determining whether these methods can at least optimize your firm's own internal workflow and information flow.

Integrate data entry and data maintenance tasks into your firm's business processes. When data creation and maintenance are completed as separate processes, the quality and completeness of the information are likely to suffer. People simply will not take the time to complete data collection and maintenance as separate tasks. Data that is not captured and maintained the first time it is created or used is less likely to be accurate and unlikely to be collected at a later date.

As you identify authoritative sources of information and increase the volume of data that you receive and transmit electronically, the integration of data collection with business processes will seem to occur naturally, but it should not be left to chance. It should be enforced with business processes and workflow rules, so that data collection and the related business process are a single, tightly integrated process.

Integrating Data Collection with Business Processes

Recently, one of the authors of this book walked into a D.C.-area bicycle shop to have a broken spoke replaced on his commuter bike. The store, part of a small chain of just two stores, had recently implemented an inventory management system that required employees to enter a Stock Keeping Unit, or SKU, for every product or service sold by the store. The SKU is a unique identifier that allows merchants to track every sales item and is typically integrated with the retail UPC barcode system described in the previous chapter. The store was in a "shake out" phase of its SKU system implementation, and the digital cash register—or point of sale (POS) computer system, as it is now called—had recently been modified so that no item could be sold to a customer without the SKU. If a store clerk would not or could not complete the inventory-control data collection task, then the business process—the sale—could not be completed.

The author had the misfortune of being the first customer to have a broken spoke replaced since the SKU system had been implemented. The $2.00 item is too small and too inconsequential to have a barcode, SKU label, or price tag attached to it. It is purchased in bulk and stored in a bin, from which repair technicians simply pluck the required quantity for each repair job. The SKU system was designed to accommodate items such as this—miscellaneous bicycle parts—by means of a loose-leaf binder containing a multipage list of miscellaneous items and their corresponding barcodes, which are scanned in lieu of a physical tag attached to the actual item. The clerk simply had to look up the item in the binder and scan the appropriate product code. But in the initial development of the inventory management system, individual spokes for this type of bike (a common repair item for sport cyclists or competitive racers but a rare one for recreational cyclists) had not been entered into the system and were not listed in the binder.

A familiar consumer experience unfolded that eventually involved four employees taking fifteen to twenty minutes to enter the $2.00 retail item (which probably cost the store much less than a dollar) into the system and sell it to the customer. Clearly, by the time the problem had been resolved, any profit on the spoke repair had been entirely consumed by the additional staff time, even with the additional revenue of the $26.25 labor charge. But, just as clearly, the store's business leaders had made a strategic decision to integrate data collection and data maintenance—inventory control—with the business process—retail sales. They very likely anticipated that these types of implementation snags would occur. They could have chosen to create a "miscellaneous repair item" SKU that would have allowed employees to complete the business process while sidestepping the inventory control system. But that would have conveyed the message to employees that the SKU system was not important, with serious long-term consequences for the integrity of the inventory control system. The business owners were willing to risk one full hour of labor (and the ire of one customer) to imbue in these employees the notion that data collection is important. It is this level of commitment to the integration of data collection and maintenance with business processes that the building industry so woefully lacks.

A common example of data that is rarely normalized in any business is so-called "personal" contact information. In the building industry, contact information may not be building information, but in many instances it can be considered vital project information. Typically, employees maintain their own contact information in their e-mail account software. As a result, vital project contact information is inconsistently compiled, is unnecessarily duplicated by many employees, is often incomplete, and is rarely available to the entire organization. Companies are rarely able to leverage the time employees spend collecting and compiling contact information to the benefit of the company. The problem is compounded by software applications that independently collect contact information, including accounting and project management applications. The most evident symptom of dysfunction is the annual exercise in many companies to assemble a comprehensive and accurate mailing list for a holiday card. If this is the situation in your firm, implementing a firm-wide customer-relationship management (CRM) system is a great way to underscore that contact information is *corporate*—and not personal—information that should be "normalized" in a centralized database (Figure 4.3). Your firm may already have the necessary software infrastructure to implement this change. The ease or difficulty with which the people in your organization adjust to this change of culture will give you valuable insight into how well your organization will adapt to a future BIM-based working environment of information sharing.

Collect all relevant information the first time. Detailed information always can be summarized, but if only summary information is initially collected, it cannot be broken down into detailed information at a later date. Individual employees may find it more expedient to gather only the information they need to get their own job done. Institute a culture of capturing all available information when it is first available; gathering it at a later date is far more costly in the long run.

The failure to collect detailed information is often proportional to the size of the company. The larger the organization, the greater the tendency among employees to gather and retain only the information they need to complete their assigned tasks. Employees must be trained to regard information holistically and appreciate that the information they create may be of value to others in their own organization, to others outside the organization on the project team, or to others both internally and externally in later stages of the building life cycle.

FIGURE 4.3
Normalized Data Structure for University Course Data.

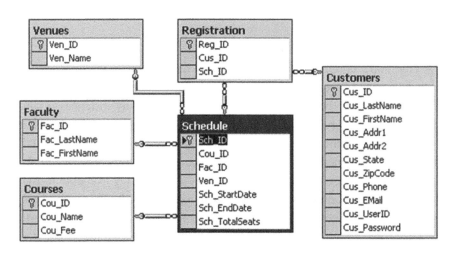

Emphasize the value of data collection and data quality. The quality of information gathered by many different people for many different purposes will vary. This is a consequence of both individual behavior and the lack of standard data formats, data collection methods, and information systems. In many cases, companies have no choice but to establish their own data quality standards. More commonly, individual employees invent their own standards on the fly. Until standard data formats, collection methods, and information systems become common, you can address this issue with a two-prong strategy. First, impress upon employees the importance of information collection and conservation habits. Second, develop a method of measuring information quality assurance—a subject we will address in greater detail in a later chapter—by requiring that all employees assign a "quality level" to every piece of information they create. (The Capability Maturity Model of the National Building Information Modeling Standard, discussed in Chapter 2, is a useful guide for establishing quality assurance metrics.) This becomes a self-regulating mechanism; no employee will want to develop a reputation for generating "low-quality" information.

Adopt open standards whenever possible. Using industry-wide open standards for interoperable information transfer will help ensure the usability and sustainability of the information—the ability to access it throughout its useful life for multiple purposes. Competitive advantages grow out of core competencies, not from proprietary data formats. Your firm's institutional business knowledge and unique skills can remain proprietary, but the format of any project information that your firm creates or compiles should conform to open standards as much as possible. Focus on providing value to your clients, not on controlling the data.

If your firm's core competency is building design, then your esoteric knowledge should be embodied in the design and not in some "creative" format for the documentation. The format of any printed documentation, for example, should conform to an open standard such as the National CAD Standard (which is itself a compilation of previous, widely accepted graphic standards), so that the contractors bidding on the construction of your design will know where to look for the information. When recipients are confident that the information conveyed to them is in a consistent format, they know where to look for specific pieces of information. As a result, they have greater confidence that they have correctly understood the full body of information conveyed to them. The result, in the case of construction documents, is tighter bids and fewer requests for information (RFIs) and change orders.

To illustrate how woefully inadequate information exchange business practices in the building industry are, consider the contrasting example of your corporate financial data. Your accounting data is organized in a standard format that your accountant can readily understand, so that he or she can correctly and quickly determine your firm's tax liability. Initially, your bookkeeper and accountant may discuss the organizational structure of your firm's chart of accounts and the format of financial reports such as the Profit/Loss Statement and Balance Sheet, but for the most part the format of your firm's financial information is based on generally accepted accounting standards, and relatively little customization is needed to organize and exchange the information.

Maintaining your accounting data in a consistent, generally accepted format assures your accountant that the information is accurate and complete. In other words, *standard data formats are a fundamental component of information assurance.* If your accounting data were organized in a proprietary, "creative" format, it might contain exactly the same information, but it might take your accountant more time to understand how it is organized than to perform the critical value-added accounting task: calculating your firm's tax liability. Despite the extra, non-value-added effort, your accountant likely would have lingering doubts about the accuracy and completeness of your financial data.

Such a lack of confidence in building information is routine for constructors during the bidding process, because the industry does not adhere to a generally accepted standard format for construction documents. Constructors compensate for the uncertainty by making assumptions or including contingencies in their bid prices. This illustrates just how serious the situation is with regard to the standardization of building data and just how easily much of it could be corrected. The core problems of the building industry have very little to do with BIM and a great deal to do with standardization of data formats and business processes. Just imagine if design professionals and constructors could share the same confidence in building data that your bookkeeper and accountant now share in your accounting data.

BUSINESS OPERATIONS AND BIM

There are five major functional areas in any company, all of which may be affected by BIM implementation:

- Marketing/Business Development
- Human Resources

- Finance
- Information Technology
- Operations

Figure 4.4 illustrates a portion of a complete business process model for an entire U.S. Federal Government department. All of the business processes for this organization, one of the largest in the world, were included in the model. If a business process model can be developed for such a mammoth governmental enterprise, surely one can be developed for every business enterprise in the building industry.

MARKETING/BUSINESS DEVELOPMENT

Depending on the size, type, and culture of the organization, marketing and business development functions are performed either by leaders with prior education and operational experience in the organization's core business, or by professional marketing and business development personnel with no direct operational experience or background, but who typically specialize in marketing one sector of the building industry and become highly knowledgeable about it. Regardless of the organizational structure and culture of the marketing function of your company, it is likely that whoever is responsible for marketing and business development has no prior operational experience with building information modeling, because the technology is so new. It is extremely important, therefore, that they acquire sufficient knowledge to market the firm's BIM expertise correctly.

Building information modeling means many things to many different people, and the expectations of clients and prospective clients may be unreasonably high. Your company needs to be able to convey accurately the scope of services and BIM expertise that it can deliver, neither overpromising nor underpromising. Marketing claims related to BIM or integrated project delivery might entail serious and costly obligations that your company may not be prepared to fulfill. Conversely, your firm may have acquired technical skills of considerable value to clients and prospective clients that distinguish your company from the competition, but that may not be readily apparent to nontechnical staff. Your marketing team needs to be able to communicate these new competencies clearly and convincingly so that significant business opportunities are not missed.

Marketing professionals will never become technical experts; it's not their job. But your professional team and your marketing team—however they are

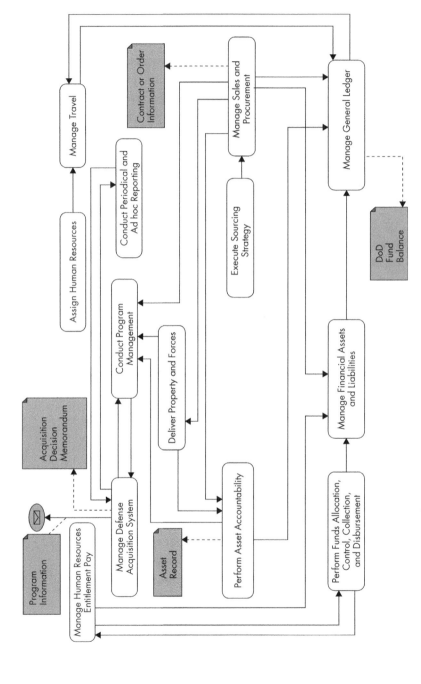

FIGURE 4.4   A Business Process Model for a Federal Government Department. Source: U.S. Department of Defense.

configured—should collaborate and communicate intensively so that the message conveyed by your marketing team is consistent, effective, and accurate. Enumerating for the marketing team the benefits of BIM that the firm can offer to prospective clients—rather than technical capabilities or functionality of BIM—is the best way to help them craft an effective message.

HUMAN RESOURCES

The demand for highly skilled design professionals is expected to outpace the supply over the next twenty years. According to the Bureau of Labor Statistics (BLS) of the U.S. Department of Labor, employment of architects and civil engineers is expected to grow by 18 percent from 2006 to 2016, and construction managers by 16 percent, faster than the average of 7–13 percent for all other occupations.[2] Over the same period, employment of lesser-skilled design professionals, which the BLS defines as "architectural and civil drafters," will grow only one-third as fast, by only 6 percent, or slower than the average for all other occupations.

This forecast is consistent with what many industry professionals have observed, particularly as BIM has become more prevalent in design firms: a clear demand trend toward a higher average level of creative and technical skill among design professionals. BIM-savvy people will be in high demand, and entry-level professionals will be expected to have a much higher level of technical knowledge. Another trend will affect the supply/demand equation for design professionals in ways that the BLS is unable to account for: as the building industry becomes increasingly integrated, design professionals are increasingly migrating to other segments of the industry, where their skills are being applied as successfully to construction or facility management as to building design.

The projected supply/demand equation will place acute pressure on the building industry to increase knowledge-worker productivity across the industry to meet the rising demand. More work will need to be done with fewer, more highly skilled people. Design professionals will be expected to have a higher level of technical skill and knowledge, while the education, knowledge, and skill of many people throughout the building industry will be expected to approach or equal that of design professionals.

In the short term, institutions of higher learning will be unable to satisfy the demand, which means employers will have to rapidly develop BIM and integrated project delivery skills internally. Organizations that choose to repeat the conventional wisdom of the CAD era, that employees will simply "pick up the skills on the job," will find themselves at a significant disadvantage. Only a strategic and methodical BIM implementation and training plan will enable companies to leverage the full business potential of BIM, for the simple reason that no individual employee can optimize the value of his or her own intelligence in a BIM-based integrated project delivery environment.

Among the strategies for transforming your organization into a highly skilled, high-value-added organization is to outsource every function not directly related to your firm's core competencies. Some or all of many support functions—IT management, Web design, reprographics services, accounting, marketing, and even certain BIM-related data collection, data entry, and data maintenance tasks—can be performed more cost effectively by third parties. Only a laserlike focus on core business operations will enable companies to foster an image in the marketplace of premier expertise.

FINANCE

There are two aspects of finance in business: accounting and financial management. Often conflated with one another, the two are not at all the same thing. The primary goal of accounting is to determine your firm's tax liability, quarterly and annually. The primary goal of financial management is to establish a profit target and to manage the firm's resources effectively to achieve it. Accounting is done by accountants and bookkeepers; financial management is the responsibility of business leaders and managers. One is a business-support function; the other is a critical business management function. Of the two, building information modeling is likely to have the greatest impact on financial management.

Let's begin by very briefly reviewing the essentials of financial management, of which there are only two core components:

- Inputs, which consist of revenue, expenses, and time (or labor)
- Outputs, which consist of typical corporate financial reports such as a profit/loss statement, a balance sheet, a cash flow report, and project job cost reports

Similarly, there are only three core activities of financial management:

- Forecasting, which consists of preparing an annual budget, an annual profit plan, and project budgets
- Measuring, which consists of analyzing the data in financial reports to determine whether budget and profit goals are being met
- Decision making, which consists of direct action to keep the entire company or individual projects on track with the forecasted financial goals

In the design and construction sectors of the building industry, effective financial management can be exceedingly challenging. It often seems to occur retroactively; the most accurate "forecasts" are made when the actual economic activity is nearly completed. Or budgets are continually revised so that they become essentially meaningless as financial management forecasting tools. As a result, financial management is not widely perceived in the building industry as a critical business management tool. This is understandable—though not necessarily defensible—in a business environment rife with ambiguity and uncertainty: projects that involve large sums of money that often span two or more fiscal years; unpredictable site conditions; volatile material, equipment and labor costs; and the vagaries of the weather. Still, the very real ambiguities and uncertainties of the industry often provide cover for even sloppier financial management, and the result is chronically low productivity and profit margins. Far too often, financial success is a result of the prodigious tactical skill—as opposed to strategic skill—of project and program managers. The heavy reliance and high value placed on tactical skill—the ability to get things done under pressure—is one reason that change is so difficult to implement in the industry. Project and program managers know that their time-tested ways of doing things work and are very reluctant to give them up for newer, untested methods.

Building information modeling, however, is very well suited to reducing ambiguity and uncertainty throughout the building design and construction process. BIM can help improve the quality and accuracy of financial forecasts, which can lead to greater productivity and profit. Building design and construction are complex endeavors that will inevitably result in mistakes, regardless of the competence and diligence of everyone involved. Mistakes, by definition, cannot be forecast, though as an industry we've become very good at developing statistical averages and planning for their inevitable occurrence by including sums for "contingencies" in project budgets. This is a tacit admission that our existing inefficient business processes have been fully optimized and that additional gains can be achieved only by implementing entirely new business processes.

BIM allows designers and constructors to construct the building virtually before it is built in reality. The benefit of BIM is not that mistakes are eliminated, but rather, to paraphrase Kimon Onuma, that project teams can discover and correct mistakes much earlier in the process, far more quickly and at a much lower cost than is otherwise possible. Resolving obvious "clashes" between building systems is only one of the most apparent benefits of BIM. Figure 4.5 illustrates this principle. Line 1 on the graph indicates the inverse relationship between time and ability of the design and construction team to affect the cost of construction; Line 2 illustrates how the cost of design changes increases as the project moves forward in the design and construction process.

FIGURE 4.5
**Design Effort and
the Cost of Change.
(Source: Patrick
MacLeamy, FAIA. Used
by permission.**

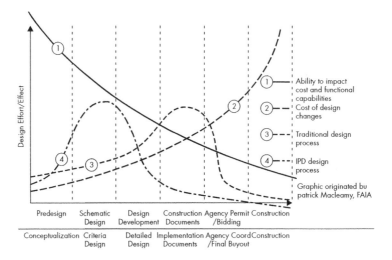

These two lines are mirror images of each other; as the ability of design and construction teams to affect the cost of a project declines, the cost of design changes increases. Therefore, teams want to make design decisions and changes as early as possible in the design and construction process. Line 3 shows how the work effort in a traditional design project peaks very late, at the point when the time/cost lines cross one another. An integrated project delivery process using BIM moves the bulk of the design effort forward, maximizing the ability of the design team to affect project cost at the lowest possible cost.

Detailed models, once completed, also can be analyzed for optimum construction sequencing and scheduling. All of these things have direct implications for greater predictability, which, in turn, allows for greatly improved financial forecasts. This alone increases the value of design and construction services to clients, who have heretofore assumed a degree of risk with respect to project cost that clients in other economic sectors would not accept from their service providers. Already, some constructors with early BIM experience have advanced the idea of "modeling the as-built condition," and then "building the as-built." In industry jargon, this is a very colorful way of saying that the end result can be known in advance; it is completely predictable. The idea that our business processes might result in a completely predictable outcome is considered radical, which only illustrates just how unpredictable our existing processes are.

INFORMATION TECHNOLOGY

In many organizations, information technology (IT) has been elevated to a strategic status on a par with revenue-generating, value-added business operations.

The appropriate status for IT is that of a vital support function, equal in importance to—but not greater than—marketing, human resources, and accounting. Like these other support functions, IT is an essential part of your organizational infrastructure—your business could not function without it. But like other support functions, IT is a cost center, not a profit center. It is a necessary commodity, not a strategic business function. Your clients expect you to have the necessary IT infrastructure to conduct business, such as e-mail, just as they expect you to have telephones and fax machines. They are no more interested in your IT infrastructure than they are in whether you have a bank account or a payroll service.

Information technology should never be equated or confused with BIM, which is—or will become—an integral part of your revenue-generating value-added business operations. Most importantly, strategic business decisions regarding business processes and workflow—such as BIM implementation—must not be subject to the approval of the IT department. No business leader would allow their HR, marketing, or accounting departments to exercise such a veto, but many grant (or cede) such authority to their IT directors, who have managed to convince their superiors that this or that business decision could have dire consequences for the firm's entire IT infrastructure. You would not let your telephone provider dictate your communications strategy; don't let your IT department dictate your business process and workflow strategy. IT professionals should be respected for the bona-fide professionals they are—just like HR professionals, marketing professionals, and accounting professionals. Their wise counsel should be taken seriously. But they should not drive the train.

Available information technology infrastructure, which remains admittedly imperfect, has nevertheless matured to the point where it can support, at a very reasonable cost, secure and robust business operations that are far more demanding than any that currently exist in the building industry. If your firm's existing IT infrastructure cannot support the BIM-related business processes and workflow that you wish to implement in your firm, then you need to ensure that your IT department has the financial, personnel, and technical resources it needs to provide your firm with the IT infrastructure your firm needs, because your competitors surely will.

OPERATIONS

Your operations are the reason your business exists. The other functional areas of your company exist to support your revenue-generating, value-added operations, which must be the primary focus of your business process reform efforts. You need total focus on this workflow challenge. Those support activities must be responsive to changes in your business operations, not the other way around.

The mantra for any BIM implementation—regardless of the business type—is workflow. Properly implemented, BIM will fundamentally change the way you do business, both internally and with business partners and clients. In the building industry, information is at the heart of workflow; it is the currency, the medium of exchange that allows everyone to do what they do. When the flow of information stops, work stops. Therefore, modifying workflow means paying close attention to information flow. The success of your BIM implementation will depend on how well you streamline your workflow and information flow, preferably by using tools such as the Business Process Modeling Notation method described in the previous chapter that allow you to revise your workflow in a methodical, documented way.

Before any company begins the workflow modeling process, however, the leaders of the company must articulate a clear strategic vision for the company. The strategic planning process must include a frank analysis of the organization's strengths and weaknesses. The goal is to determine how the organization fits into the industry as a whole, its specific markets, and the prevailing local, regional, or national business culture. No single person or organization can possess all of the talent, skill, and resources to provide all of the products and services in the building industry. Business leaders understand this implicitly, of course, but often fail to articulate it explicitly as part of their business strategy. Collaboration and cooperation with others is vital to the success of individual companies. Businesses add little value—and erode their own profitability—by duplicating what others may be able to do better. Everyone has a niche; success and profit are achieved by exploiting that niche.

Advocating that businesses engage in strategic planning is conventional business-school wisdom. But for business organizations in the building industry, building information modeling heightens the importance, value, and necessity of strategic planning. A technology designed to support radically new business processes and relationships cannot be implemented successfully in a haphazard manner.

As an organization embarks on a BIM implementation, it must determine as part of its strategic planning process which skills the organization will acquire internally and which skills will be acquired through business relationships (consultants and subcontractors). Leveraging the talent, skill, and resources of others without acquiring those assets directly can vastly improve an organization's profitability and nimbleness—its ability to respond quickly to the demands of the marketplace. In the CAD era, many design firms—and some building owners—made substantial, early investments in proprietary CAD technology for design that was soon surpassed by better technology. These "early adopter" firms soon found themselves caught in a trap of obsolescence, forced to recoup a return on their initial investments before they could adopt newer, cheaper, and better technologies. To a large degree, the risk of obsolescence could not be avoided, as hardware costs were relatively high and most software was available only as a workstation product on a per-seat-license basis and required considerable training to master. Today, computer hardware is far less costly and software can be acquired through a number of far more flexible purchase options: flexible enterprise licensing; concurrent user licensing; or Web-based software as a service, which can help minimize an organization's IT infrastructure needs and enable companies to acquire technology without incurring the significant upfront cost of shrink-wrapped software.

The building industry also has a strong track record with respect to distributed resources and expertise. In the construction sector, though many aspects of project execution may be inefficient, the basic organizational structure of the industry is actually highly efficient in several important ways. Most prime contractor construction companies, regardless of size and scale, have a very high net revenue per employee. The dollar value of heavy construction equipment that these companies own is very low in comparison to their business volume. They typically outsource most of the construction services they provide to subcontractors and lease or rent heavy equipment as needed on a project-by-project basis (or depend on their subcontractors to provide it). Construction firms of all sizes—from single-family home builders to multi-national giants—generally fit this business profile, which allows the construction industry as a whole to maintain relatively low overhead and to move labor and equipment from project to project and region to region very efficiently. The "system" is self-regulating; within the limits of overall economic performance, the degree of disparity, if any, between supply and demand of equipment and labor is fairly constant nationwide.

The same distributed business model can be applied to the creation and maintenance of building information, and to the resources needed for building information modeling. The architect Frank Gehry, renowned for designing some of the most technically complex building forms in the world, readily admits his near-total incompetence with computer technology. But he recognized long ago that his vision could not be realized by conventional methods and began acquiring the appropriate resources—people, equipment and software—to meet his business needs. He did not follow a conventional path, acquiring the most popular technology. Instead, he considered his business model—his "core competency" in designing complex building forms—then set about acquiring the talent and technology that would allow him to fulfill his vision. His "BIM implementation strategy" eventually led him to establish a new company, Gehry Technologies, to commercialize the technical expertise that his design firm had developed. Today, Gehry's architecture and technology companies are widely regarded as worldwide leaders in advanced building design technology, and many other firms in both design and construction now benefit from Gehry's innovations. And yet, by his own admission, Gehry's personal competency in even basic computer technology has hardly advanced at all.

We simply cannot underscore enough for business leaders the importance of aligning their technology strategies with their organization's core competencies. It is a far, far different strategy than simply "getting BIM" and "doing BIM." The information modeling needs of design professionals, constructors, owners, and operators are completely different from one another, as are the appropriate tools. The common goals across all business types are:

- Get people to work together as a team;
- Improve communication;
- Perform more tasks in parallel and fewer tasks in sequence.

Your effectiveness in achieving these goals is what will set your firm apart from your competitors.

WORKFLOW VISUALIZATION

Once an organization has established a clear strategic vision, its business leaders can begin to develop a more detailed vision of how the organization will function in a BIM-based, integrated project delivery environment. This is not unlike any design project. What are the requirements? What is the desired outcome? What will the new operational model look like? Business process modeling tools and creative thinking must be applied toward the design of the company. Urban areas that do not plan for growth end up tangled in inefficient sprawl. Likewise, a company whose growth and development is unplanned is likely to be inefficient and ineffective.

Based on the early experience with BIM implementation among design and construction firms, we believe that the core component of an optimum, BIM-based organizational business model is the integrated, cross-functional, co-located project team, equipped with robust, room-size video conferencing and Internet communication capabilities, preferably located in close proximity to convenient transportation infrastructure such as an airport. Proximity of the project team to the project site may or may not be important, depending on the project.

The co-located, communications-equipped environment fosters frequent, informal, face-to-face, and remote meetings and rapid decision making. Obviously, it is not possible for every individual in every organization involved in a project to be co-located. The co-located team consists of the senior team members in each organization who have the authority to make decisions on behalf of their organizations, supported by their own project teams located within their own organizations. The communications infrastructure—including video and digital information exchange and collaboration capabilities—should be sufficiently robust and available so as to make physical distance between extended team members seem as inconsequential as possible.

Whenever possible, the project team environment should be established as early as possible in the life cycle of the project and should consist of all major design and construction team members, including the owner, major subcontractors, and fabricators. The constellation of co-located project team members may change over time with the intensity of individual responsibilities. Ideally, stewardship of the building information created by the project team is shared, though each team member, by virtue of their physical presence, retains the ability to exercise responsible control of the information within their assigned realm of responsibility and expertise, and can still be recognized as the sole authoritative source and authority of that information.

Many concerns regarding liability of authorship appear to diminish in this type of collaborative, real-time, consensus-based, decision-making environment. A great deal of correspondence designed to limit the liability of one party by putting another party "on notice" with copies to every other party "for their information" is eliminated. Much of this type of correspondence is intended to create a defensive evidentiary record in anticipation of some unknown and undefined future legal action. It is difficult to explain to seasoned professionals with no prior integrated project delivery experience, but the intensive interpersonal interaction of co-located project teams appears to mitigate such risks by reducing the perceived need for defensive documentation. Many legal disputes in the industry arise when adversarial parties develop a different understanding of the same set of facts. When you are looking someone in the eye and talking to that person every day, you have greater confidence that you share a mutual understanding, and you worry less about what the other might be thinking or what he or she might say or do in your absence. Because the project is developed collaboratively, there is also a greater sense of joint responsibility for errors and omissions, which fosters a greater sense of mutual responsibility for correcting them promptly, in lieu of the current reflexive response to point the finger of blame at a particular team member.

The degree to which this type of working environment can be established will be predicated on many factors, not the least of which is the contractual project delivery mechanism for the project. It is, however, possible to achieve at least some of this type of collaborative environment on every project. There is nothing about our existing project delivery methods that prohibit more intensive team collaboration.

The emerging co-located project team environment is only one of many possible scenarios for BIM-based integrated project delivery. Others may work equally well, and we encourage readers to experiment. Because so little building information modeling data exists to date, business models for later phases of the building life cycle—and for integration of workflows between phases—are harder to come by. But it is reasonable to assume that the organizational model for facility management, maintenance, and operations will develop similar collaborative characteristics, with distinct functional areas coalescing into cross-functional, integrated teams.

ENDNOTES

1. Wikipedia, "Database Normalization," Wikipedia, http://en.wikipedia.org/wiki/Database_normalization.
2. Bureau of Labor Statistics, "Occupational Outlook Handbook, 2008–09 Edition," U.S. Department of Labor, http://www.bls.gov/oco/.

CHAPTER 5

The Building Life Cycle

What gets us into trouble is not what we don't know, but what we know for sure that just ain't so.

—Mark Twain

The life cycle aspect of building information modeling is primarily what sets it apart from preceding digital technologies, which were designed to automate specific tasks in specific phases of the building life cycle for specific sectors of the building industry, such as design, construction, and facility management. In preceding chapters, we have described how the detailed, holistic, graphically expressive, and data-rich nature of BIM fosters an environment for greater collaboration during the design and construction phases of buildings. The highly structured building information generated during design and construction will similarly create an opportunity for maintaining and using building information throughout the building life cycle (see Figure 5.1).

LIFE CYCLE VIEWS OF BUILDING INFORMATION

There are fewer signposts or benchmarks to support the holistic view spanning the entire life cycle of facilities. The paucity of case studies, however, is not surprising. The volume of building information that has been created thus far

FIGURE 5.1
The Building
Information Life Cycle.
(Source:
buildingSMART™. Used
by permission.)

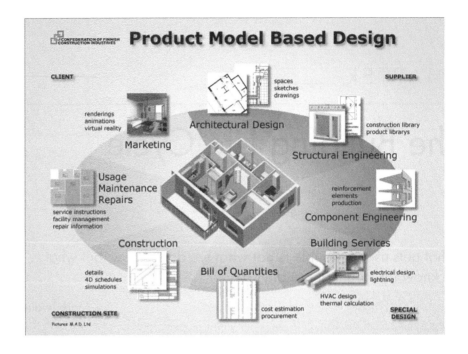

is relatively small, and only a fraction of the BIM data produced to date has been passed on to building owners and facility managers, many of whom lack the tools and organizational infrastructure to use the data effectively. Another factor is that while an increasing number of design and construction professionals are showing a willingness to share information with one another, there is a greater reluctance on the part of design professionals to pass building design information onto owners or other third parties for business processes that extend far into the future, for which the information was not originally intended, and over which they will have no control. Statutes of repose for professional liability vary from state to state, but in many cases the "tail" of liability for design professionals extends many years after they have fulfilled their contractual obligations, often far beyond the period of the much lesser liability of constructors. Many design professionals reasonably conclude that the risks of releasing digital building information are simply too great.

This is one of the dilemmas of the building industry that is seemingly unsolvable, short of modifying state laws nationwide to narrow the scope of professional liability for design professionals. But as the emerging culture of integrated project delivery demonstrates, conventional, seemingly intractable problems are not likely to be solved with conventional solutions. The important lesson of integrated project delivery that emerges from the preceding chapter is that intensive, *interpersonal collaboration,* in which each participant can at least observe how others use building information, *is a vital component of information assurance.* The lack of similar oversight or involvement by downstream users hampers the transfer of information at the completion of the design and construction phase. The solution, then, may be to develop similar information oversight mechanisms.

Some way must be found to assure information authors that the information they create will be used appropriately—and at no additional risk to them— once the information is no longer within their realm of responsible control (see Figure 5.2). Once again, the building industry can benefit by looking to other industries that have overcome similar challenges. A useful example can be found in the package shipping industry. Before package-tracking technology developed, anyone shipping a package simply had to accept that a package would disappear into the package-shipping industrial complex, never to be seen again until it arrived at the doorstep of the recipient. The status of packages in transit was simply too difficult to track. In the late twentieth century, shipping companies began to develop technology for tracking packages, largely for their own internal use. One enterprising company, soon followed by competitors, decided to provide that internal tracking information to customers by telephone. This new customer service became wildly popular but soon ballooned out of control, overwhelming the shipping companies' customer service operations and dramatically increasing customer service costs. Shippers quickly shifted from human customer support to electronic voice package tracking

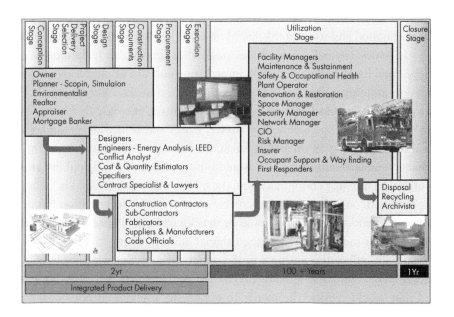

**FIGURE 5.2
Building Information
Stakeholders.**

support, but customer service infrastructure support costs remained high. Having raised customers' expectations for information, they were unable to turn back the clock. Customers expected to be able to call a toll-free number twenty-four hours a day, seven days a week, to inquire about the status of their packages. Many customers made such calls three or four times for every package.

With the advent of the Internet, shipping companies soon realized that they could move their tracking systems into a Web-based, customer-accessible environment, and the customer-service nightmare soon ended. Customers could track their own packages by logging into the same tracking system used by the shipping companies themselves, another example of the tight integration of data collection and maintenance with business processes. Customer service representatives, human or electronic, were no longer needed, and toll-free telephone customer support all but disappeared.

As shipping customers gained greater and greater access to package status information, a curious thing happened: complaints from customers about on-time arrival of packages declined, regardless of the actual on-time delivery rates. What mattered to most customers was not the exact day or time a package would be received; instead, customers placed higher value on knowing the status of their package in transit, whatever that status might be. Nothing about the core value-added service—package delivery—had changed. Customers still had to give up "responsible control" over their packages. Once they selected their delivery method, they were powerless to affect the status of the package between the moment it was shipped and the moment it arrived. But they gained "package status control," which turned out to be the information assurance that most customers wanted: knowledge that their package was making steady progress toward its destination and was not misplaced or idling in some vast remote warehouse.

It is not beyond the realm of possibility for similar technologies to be developed for BIM data, in which information authors would receive the information assurance they most desire: knowledge that the information they created is not being used or modified in a way that will increase their professional liability. Much of the technology already exists and is in widespread commercial use, including automatic notification, data file modification tracking, and change-management logging. It would not be difficult for original information authors to monitor continually how the information they created is being used or modified long after their responsible control of that information has ended. BIM audit and analysis tools could be developed to query and analyze building information models and alert the original authors of any changes that might potentially affect the author's professional liability. Software tools could be developed, contractual agreements drafted, and business processes implemented that would require the owner to obtain the original author's consent (or their author's corporate heirs and assigns) before proceeding with certain types of changes to the model, or indemnify the original author for changes that, in the author's opinion, would unduly expose the author to increased risk.

From the standpoint of our current business culture, such a scenario may seem hopelessly optimistic or naïve. However, another consequence of BIM and integrated project delivery is a shift from a project-to-project business paradigm and toward long-term business relationships with valued partners. In such an environment, the "life cycle information assurance" scenario described above becomes far more plausible.

It is likely, also, that a new building industry profession or service—building information stewardship—will emerge to meet market demand to sustain building information over the life of a building. In addition to the many people who create or handle building information throughout the life of a building, there are many potential customers for accurate, up-to-date building information that could be extracted from building information models, including emergency responders, insurers, real property portfolio (investment) managers, real estate brokers, and prospective purchasers (future owners) of buildings. Accurate, up-to-date building information, as we've noted previously, will become a tangible asset in its own right. It is not much of a leap to contemplate that a business opportunity will emerge for "building information assurance agents" who will not only maintain building information models, but will also certify that the information they convey to third parties—for a price—accurately reflect real-world conditions.

THE FEASIBILITY, PLANNING, AND DEVELOPMENT VIEW

We tend to think of building design as the start of the building life cycle, but, in fact, the life cycle of a building begins long before the design process begins—sometimes many years before. Much of the information created in the feasibility, planning, and early development phase of a building is building information. This is the first opportunity to begin compiling information about the building. The data may be a pro forma, a table of functional space requirements, a budget, or a construction cost estimate.

The most important thing that professionals in this phase of the building life cycle can do is recognize that the information they create is building information, and take the simple and easy step of compiling and conserving it for possible later use. Many of the tasks performed during this phase in the building life cycle rely on spreadsheets for alphanumeric building data. *Any spreadsheet of building information, because it is a form of structured information, is a building information model.*[1] The second most important thing that can be done in this phase of the building life cycle is to make a concerted effort to compile building information in spreadsheets instead of in text documents whenever possible. A spreadsheet is a form of structured information; a text document is not. Obviously, this will not be possible for reports and other narrative documents, but even in these instances, any tabular data included in text documents should first be compiled as spreadsheets and subsequently embedded in the text documents with external reference links.

All other things being equal, compiling data in spreadsheets instead of in text documents should have no effect on workflow, require no additional training, and require no additional investment in software or other technology. The potential benefit, however, is enormous. The structured data contained in spreadsheets can be imported directly into BIM programming and planning tools such as the Onuma Planning System (OPS) or Trelligence Affinity. To optimize integration with these tools, the format of spreadsheet data (column headings) may have to be determined in advance or modified prior to import, but the critical factor is that the data exists in a spreadsheet format. Forward-thinking professionals in this phase of the building life cycle, without any prior knowledge of how the information they create might be used and by whom, can begin the building information life cycle simply by following these few recommendations. This is the ultimate "low-hanging fruit"—nothing more low-tech has as much high-tech potential.

As data generated in this phase becomes more accessible to BIM planning and programming tools, we can anticipate that new BIM tools will emerge to reach back and support feasibility and planning processes. It is also likely that this sector of the industry will be the object of a data standardization and normalization effort to support streamlined workflow, decision making, and information flow to information stewards in later phases. Better information creates markets for better data analysis tools, which spawns still better information. BIM design tools are already being used for scenario planning, allowing designers to test many possible design scenarios against many possible criteria, an exercise that until now has been cost prohibitive. It is not difficult to imagine the rapid emergence of sophisticated scenario planning tools in predesign phases. It would be relatively easy, for example, to leverage statistical economic data for planning purposes and to subject alphanumeric planning information to customized rules-based analysis.

THE DESIGN AND CONSTRUCTION VIEW

Much of the focus of this book—and the initial focus of BIM and integrated project delivery—is on the design and construction phase of buildings. Though it is neither the most important nor the most costly phase in the life cycle of a building, it is in this phase that the largest volume of BIM data is being created and where the most advanced BIM experts are concentrated. By default, design and construction is driving BIM adoption and innovation—and business process reform—throughout the building industry. This provides an opportunity for enterprising individuals and companies to apply their expertise and extend their service offerings both backward and forward to other phases of the building life cycle.

Though innovation may be slow in individual companies or markets, BIM experts in the design and construction sector—commonly known by the acronym AEC, or architecture, engineering, and construction—are perhaps more mindful of it than their colleagues in other sectors of the life cycle view. Among the foundational standardization efforts of the buildingSMART alliance and its worldwide counterparts are the Information Delivery Manuals (IDMs) and Model View Definitions (MVDs). These are examples of the AEC sector's collective recognition that better information is needed to spur the development of better tools, and tacit acknowledgment the AEC sector has utterly failed to document its own existing business processes properly.

As the International Code Council (ICC) has developed its SmartCodes technology for automated code checking, David Conover, ICC's former senior advisor, has observed that a typical building information model may contain 100 percent of the information needed to convey design intent under the currently prevailing standard of care, but contains only 85 percent of the information needed for automated code checking (see Figure 5.3). This is both good news— the vast majority of the needed information is already there—and bad news— adding the additional 15 percent will require time and money, and no one is yet certain, in every instance, of the precise content to be included in the missing 15 percent. Foundational technologies such as IDM and MVD will help identify exactly what that information is by defining, for example, a model definition view for automated code checking and the information that must be included to

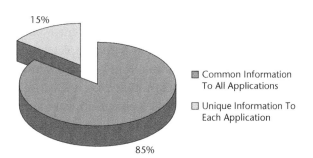

15%

85%

■ Common Information To All Applications

□ Unique Information To Each Application

**FIGURE 5.3
A Typical Building
Information Model.**

generate that view. This will create business opportunities and reduce ambiguity and risk for both service providers and clients by enabling them to align their expectations with respect to the scope of services.

Building owners will be able to choose the types of analysis that they wish to have performed on their building information models. Each type of analysis may add an incremental cost, because more information—the missing 15 percent—may need to be collected for each type of analysis. Some types of analysis, such as clash detection, very likely will be expected to be included in the base level of service, since resolving physical interference issues has long been within the professional standard of care for building design professionals. Other types of analysis, such as additional development of the model for automated code-compliance checking, energy consumption and building performance, sustainability, LEED certification compliance, and many others, might be performed on an elective and value-added basis. In each case, owners can be presented with clear choices: pay an additional fee to add the information needed for automated code checking, for example, or submit conventional construction documents for conventional code review. If time is of the essence, as it frequently is, the added value of enriching the model for code checking should be very easy for the owner and the designer to quantify. With the needed data sets for each type of analysis sufficiently defined by IDMs and MVDs, building owners and design professionals will be able to negotiate the scope of services from a list of many possible analyses. Over time, as the BIM knowledge and skill of design professionals increase and the capture of more detailed information in an integrated project delivery environment becomes routine, the volume of "additional" information that will be needed to complete multiple analyses will decline, as will the cost of enriching the models.

THE OPERATIONS AND MAINTENANCE VIEW

Throughout the life of a facility, the basic information about the facility is collected over and over again for all sorts of reasons. Very little of the large body of information created during the design and construction phase is transferred to facility managers in any consistent, methodical way. The business process for this vital information transfer is completely undefined. Most facility managers are unable to indicate clearly what information they need and lack the information management tools to use it effectively. Most design and construction professionals haven't a clue as to what information they should provide facility managers and do not yet perceive the information they create as a value-added product or service.

Much of the information created in the design and construction phase is not useful for facility management purposes, while much of the information needed for facility management purposes is not created or properly compiled in the design and construction phase. Frequently, the problem of information transfer is one of semantics: the model may contain detailed information about the area of a building, for example, but designers and constructors may define "area" completely differently than facility managers. Because the two parties do not understand each other's definitions, they lack *information assurance* at the point of information transfer. The result is that facility managers routinely re-create area data, since that is the only way to assure themselves that the data is correct for facility management purposes.

The solution, once again, is a model view definition for facility management that is published and widely understood by all parties. Then, if any information is missing from the facility management MVD, the problem becomes one of filling in the missing information, rather than re-creating the entire data set from scratch. Service providers and clients then will be able to agree on an appropriate fee for this value-added service, although ideally, as data standards mature, the data needed for this view will be a natural outcome or by-product of earlier processes in the building life cycle.

If facility managers receive true as-built information from designers and constructors and have confidence in that information, they will be able to increase the efficiency of their facility management operations, save significant amounts of money over the life of the facility, and perhaps even extend the useful life of the facility. While many tax amortization and longevity models only project a useful life for a typical building of twenty to seventy years, the reality is that a well-designed and sustainable facility can last much longer. This fact came home for one of the authors on a recent trip to Paris that included a dinner at a restaurant that had been in continuous operation since the 1400s. The operations and maintenance record of that facility, if it existed, would tell a remarkable story of thousands of renovation and maintenance projects; building, fire, and health code upgrades; equipment repair and replacement; fuel upgrades from wood to coal to gas, oil, or electricity for cooking and heating; lighting upgrades from candles to oil lamps to gas lamps to electricity; plumbing upgrades from outhouses and hand-dug wells to indoor plumbing; continuous repair and maintenance of the roof and exterior building envelope; and, finally, the introduction of air conditioning. Few of the changes that this facility has endured over six hundred years could have been anticipated, and yet it has proven sufficiently adaptable to continue serving its originally intended purpose.

While few buildings designed today can be expected to be used for the same purpose for such a long period of time, a building information model can

FIGURE 5.4
BIM Models That
Accurately Represent
As-Built Conditions.
(Source: M.A. Mortensen
Company and the Uni-
versity of Washington.
Used by permission.)

help ensure that every facility remains in use for one purpose or another for as long as possible, by allowing building owners and facility managers to conduct detailed analyses not unlike those that have emerged in the building design and construction phase. Many possible renovation and adaptive reuse scenarios, for example, could be generated rapidly and inexpensively for buildings that are no longer suitable or needed for their originally intended purpose. Those scenarios could include an analysis of the building materials and equipment that would have to be removed from the building for each scenario and how those materials could be removed most cost-effectively. The number of criteria that could be considered in the scenario planning process is virtually limitless, provided the right data is available (see Figure 5.4).

The actual operations and maintenance of the facility could be greatly streamlined and improved with a facility management model that includes comprehensive warranty information, routine equipment maintenance information, the estimated useful life of major building components, and much more. Facility management models could be integrated with building systems controls for monitoring building performance in real time, enabling facility managers to fine-tune building equipment for optimal performance. The cost of operating and maintaining facilities would become far more predictable, which would improve their financial performance as investment instruments.

While the transfer of information to facility managers at the end of design and construction remains a challenge, some building owners are choosing not to wait and have begun building as-built models of their existing facilities. The U.S. Coast Guard, among others, has begun creating models for asset management purposes and is experimenting with using BIM for operations and maintenance.

The most visible example of a BIM created for an existing building is the model created for the Sydney Opera House in Sydney, Australia (see Figure 5.5).

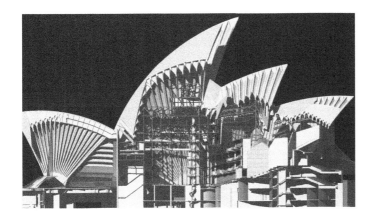

FIGURE 5.5
Section through the BIM for the Sydney Opera House.
Section through Existing Opera Theatre, © copyright Sydney Opera House.

At a professional conference at which the project was presented, Stuart Bull, BIM coordinator in the Sydney office of the international design firm ARUP, remarked that it was the first time in fifty years that the owner and its consulting design professionals truly understood how the building functioned. The model was used to conduct acoustic analysis, the results of which were used to fine-tune the opera house itself by adjusting the sound-reflecting baffles to ensure optimum sound quality. Just as some constructors are finding it profitable to create models for construction purposes when models are not provided by design professionals, in certain instances building owners will recognize significant financial benefits to creating as-built models of existing buildings. These cases, while still rare, will increase the total volume of BIM data, which, hopefully, will be incorporated into the continuing life cycle for these buildings.

Building information modeling has vast implications for the facility management phase of buildings. One of the authors had the opportunity in a recent employment position to develop the concept design for the logical and physical assets of a government infrastructure operations center (see Figure 5.6). The physical facility was designed and built to house a sophisticated IT infrastructure for a government facility that housed national security assets. It was expected to operate continuously, 24 hours a day, 7 days a week, 365 days a year. Downtime was not an option. For the design team, the primary responsibility was the IT infrastructure, not the physical facility, but the performance specifications demanded a very high level of reliability, which was often negatively affected by the facility. One night prior to installation of the new system, the computer servers automatically notified the server team of system failure, which prompted an emergency response of IT personnel to return the system to operational status (Figure 5.7 illustrates the resulting dip in system performance.)

FIGURE 5.6
Federal Agency
Operations Center.

Until the team arrived on site, the cause of the failure was unknown. Upon arrival they discovered that the air conditioning unit for the space in which the servers were housed had failed. The servers were designed to monitor system performance and recognize a system crash but not to detect an underlying, external cause such as a rise in temperature.

The new operations center was equipped with building systems monitoring controls, which could alert facility managers to an abnormal rise in temperature long before the temperature had risen to the point where it could cause computer servers to crash. Most importantly, because the exact cause of the problem could be detected, the correct responders—HVAC technicians—could be dispatched to fix the problem before it became an IT problem. In addition to temperature, the new system also had sensors for humidity, fire, smoke, water, and poisonous gas. The reliability of the IT infrastructure rose from 99 percent to 99.99 percent (see Figure 5.7). That difference may seem inconsequential, but it is the difference between 87 hours and 36 minutes versus less than 53 minutes of downtime *per year.* Some systems actually achieved 100 percent operational time. Facility-related causes of downtime were virtually eliminated.

It took a significant amount of time, money, and effort to achieve these results. It would have been far easier and less expensive if an accurate model of the facility had existed, and the design team had accurate information about the operating tolerances for all existing equipment. In addition, information about the expected useful life of the equipment, the expected mean time

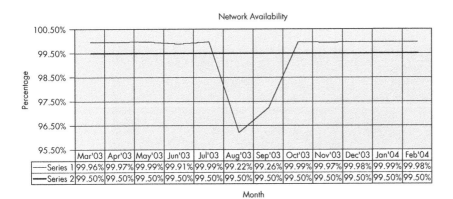

between installation and failure, and the parts needed for routine mainte-nance—information that existed in some form, somewhere, at some time—would have been tremendously valuable. A proposed solution for this problem, the Construction Operations Building Information Exchange (COBIE), is described in Chapter 7.

As building systems become more complex to meet rising energy standards and other performance criteria, operational information about facilities will become ever more important. Most of the controls and monitoring devices needed for highly sophisticated monitoring are already commercially available for a reasonable cost. The raw data generated by these sensors, however, are of value only if they can be analyzed and measured against the original design performance specifications. Not having to re-collect that design information during the operations phase of buildings would make implementation and use of monitoring systems much more cost effective.

THE OWNERSHIP AND ASSET MANAGEMENT VIEW

The building life cycle begins and ends with alphanumeric data. The final data set is concerned with managing buildings and their most valuable contents as financial assets. Buildings—large, tangible, immovable objects—are increas-ingly being monetized through financial instruments such as real estate invest-ment trusts (REITs) and managed as monetary assets with real asset portfolio management tools, which are a type of financial management tool. There is an overlap in the data set needed for building operations and maintenance and real asset management, but managing buildings as assets is more concerned with leveraging the assets for the greatest possible return, whether for space utiliza-tion or income, than for keeping the place running.

FIGURE 5.8 Onuma Planning System Virtual Real Estate Portfolio. (Source: ONUMA Inc. Used by permission.)

Real property owners with even modest real estate portfolios often have difficulty keeping track of their property assets. For large property owners, the challenges can reach comical proportions. For *very* large property owners such as the U.S. government, it can be almost unmanageable. Very few property owners are able to say with any confidence exactly how many square feet of building they own.

Property owners have long maintained tabular data in their real asset portfolios, but comparisons between the data and the "facts on the ground" often reveal wide disparities in such vital data as net leasable area, occupancy, and even tenancy. There have been many reported instances of spatial data verification that revealed significant differences between the volume of space being leased and the actual space occupied. Building information models that accurately represent real-world conditions in an owner-friendly user interface can help close this information gap. Figure 5.8 shows a "visual real estate portfolio" developed using the Onuma Planning System; it provides both graphical and detailed alphanumeric information about real property assets in a simple, easy navigated interface.

The information gap stems from the fact that unless building information and business processes are tightly integrated, it is almost impossible to sustain the validity of real asset data for any reasonable length of time. Several years ago, the U.S. Coast Guard, with full awareness of the scope of their asset management challenges, decided to begin using BIM concepts to validate and maintain real asset data. They started with what they called "BIM blobs," digital objects that represented buildings and other immovable shore assets and whose only attributes were a unique identifier for each building—definitive proof that the building exists—and its approximate total area. This information was not much more useful than the tabular data it replaced, but it demonstrated the Coast Guard's early commitment to compiling building data in a structured, object-based environment, and a strategic bet that such an environment could eventually house a far richer, more useful building information database.

Over time, as the need has arisen or funds have become available, the models were enriched with additional information such as the basic geometry of buildings, many of which have since been geospatially located so that the relative relationships between buildings could be definitively documented (see Figure 5.9). Some Coast Guard installations have added maintenance information to their models and can now accurately predict how funds should optimally be used to ensure that operations can be sustained to the highest possible level. Being able to express the data visually and geometrically also enables decision makers to develop a quick understanding of the relative condition of various installations.

FIGURE 5.9
Onuma Planning
System BIMBlobs at
Four Levels of Maturity.
(Source: ONUMA Inc.
Used by permission.)

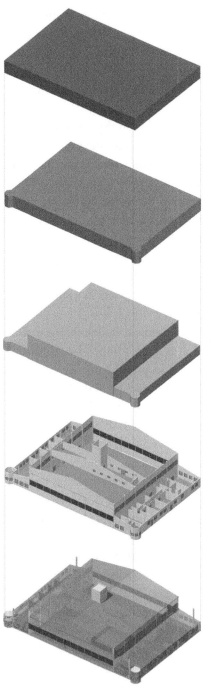

Simple Mass Defining
Total Square Footage

Mass With Rough
Outline Defining Total
Square Footage

Mass With Rough
Outline Defining Total
Square Footage and
Height

Maturing Model
Identifying Spaces

Maturing Model With
Spaces, Other
Elements and Data

The object lessons of the Coast Guard experience is that the move toward structured building information throughout the building life cycle can begin at any point along the life cycle and that the available information does not have to be complete or fully accurate for the information compilation process to begin. As long as workflows and business processes are in place to gather, store, and validate the quality of the information compiled, even the barest outline of an information database can be useful, and can provide a framework for methodically gathering more detailed information over time. Metadata is so important because, like a library card catalog, it allows you to compile and quickly retrieve large volumes of information even if the information beyond the metadata level is in unstructured form.

 ENDNOTES

1. As we have previously, we credit Kimon Onuma, FAIA, for this pioneering insight.

CHAPTER 6

Building Information Exchange Challenges

The most successful people in life are generally those with the best information.

—Benjamin Disraeli

The value of building information is directly proportional to how accurately it reflects real-world conditions and to the ease with which it can be compiled from authoritative sources and transferred from one information steward to another throughout the life cycle of a building. In order to be reliable and useful to anyone, building information must be accurate, accessible, complete, and transparent. It must be safeguarded through proper access controls at all times, to ensure that the information is the work of responsible, authorized parties and to protect against its unauthorized access or misuse.

The building industry is continually engaged in an endless cycle of forensic information analysis and data collection. Building information today is so unreliable that the cost of data collection is embedded in almost every business process in every phase of the life cycle of every building. The accuracy and reliability of building information degrade rapidly and early in the building life cycle. The chain of custody of building information breaks down even before construction of the building begins. Because it is so rarely compiled, preserved, and maintained in a consistent and methodical manner, the validity of most building information becomes suspect almost immediately after it is created.

It must be independently and repeatedly confirmed at every stage and for every task. Though information from previous phases of a building's life cycle may be readily available, doubts about its accuracy and provenance lead responsible professionals to disregard it in favor of collecting the data again, either by field inspection of the actual building or by contacting the original authoritative source, such as the manufacturer of an installed building product or system.

This repetitive, costly, and laborious data collection could be eliminated or significantly reduced if building information were compiled and maintained in such a way as to be considered accurate and reliable by anyone who has a need to access and use it. For the most part, these obstacles cannot be overcome with technology. The primary obstacles are of a logistical and not technical nature, and can only be resolved with business solutions: industry-wide data definition and information exchange standards. Software companies bear some responsibility for not making it easier for us to exchange electronic information easily, but it is not the responsibility of software companies to define our business cases—what information needs to be exchanged when, and how it should be organized. We need to get our own house in order, so that we can articulate our information exchange requirements more clearly to our software providers. We then need to make it clear to software providers that seamless and reliable information exchange is a priority to the building industry. This chapter is devoted to describing the scope of information exchange problems that we must address as an industry. Chapter 7 describes some of the efforts being made to address them.

Data, Information, Knowledge, and Wisdom

The dictionary definitions for data and information are nearly identical, but for purposes of building information modeling, we draw a clear distinction between the two and place them on a continuum of data, information, knowledge, and wisdom. Developing an understanding of this continuum is essential to understanding the value of building information modeling and using building information effectively.

Data is a collection of empirical observations or facts, *especially when organized for subsequent analysis.*[1] For example, if you place a measured rule in a vertical position on the bank of a river and make a series of observations as the river rises and overflows its banks, the written record of your observations is *data.*

Information is a collection of facts or observations *that are used to reason, make decisions, or from which conclusions may be drawn.*[2] No conclusions were drawn from recording the rising river waters, but after the floodwaters have receded, you observe that homes in which the heating and cooling equipment were installed above the first floor level are returned to habitable condition more quickly and at a lower cost than homes in which the same equipment is located in the basement. You have drawn conclusions from *information*.

Knowledge is familiarity, awareness, or understanding gained through experience or study.[3] As a result of the information acquired about the effect of floodwater on habitable structures, you propose that the building code be amended to require that new or renovated buildings in the floodplain be designed with heating and cooling systems at the second floor level or above. The application of information to a new situation from observations of an earlier situation demonstrates *knowledge*.

Wisdom is the ability to discern or judge what is true, right, or lasting; it is manifested as insight, common sense, and good judgment.[4] Your knowledge of the effect of floodwaters on habitable structures leads you not to design your own home differently, but to choose a building site well above the floodplain. Your good judgment is a reflection of *wisdom*.

Building information modeling is nothing more than a mechanism to transform *data* into *information* to gain the *knowledge* that allows us to act with *wisdom*. We begin the process by compiling *data* into a *collection of facts* that give us the *awareness* we need to exercise *good judgment*.

1. *The American Heritage Dictionary of the English Language*, 4th ed. (Houghton Mifflin Company, 2004).
2. Dictionary.com, "Wordnet 3.0," Princeton University, http://dictionary.reference.com/browse/information.
3. *American Heritage Dictionary*.
4. Ibid.

INFORMATION MANAGEMENT

As the volume of available digital building information grows, even for a single building, it has the potential for becoming unmanageable. Methods need to be deployed to find and retrieve the right information quickly. One of the available tools is *metadata,* which is nothing more than data about data.

The cards in a library card catalog are a form of metadata. The information contained on each individual card is in a highly structured standard format that allows any library patron to retrieve any book in any library in a matter of minutes, whether the collection consists of ten, 10,000, or 10 million volumes (see Figure 6.1). Because the information in library card catalogs was already so highly structured and uniform from one library to another, the transition to digital card catalogs was relatively easy, and the conversion made the metadata fully searchable, further reducing the amount of time needed to conduct library research. The label on a soup can is another form of metadata. While it contains nutritional information, a list of ingredients, and a product bar code, the most important piece of information on the label is the type of soup inside the can; you don't need to open every can to learn which one contains cream of mushroom or chicken noodle. The digital data of most building industry organizations is a world of cans without labels, a library of books stored randomly and uncatalogued. Imagine a library in which you have to open each book before you could know what the book is about; more often than not, this is the state of "information management" in the building industry. As a result, we waste a tremendous amount of time looking for information, validating information, and re-creating information because either we don't have confidence

FIGURE 6.1
Card Catalog.
Flicker Commons. Photo
by Sage Ross.

in the information that we already have in our hands, or we simply can't find it.

The problem can be solved for some organizations by deploying sophisticated data indexing and search tools such as Newforma Project Center, but a large part of the answer is organizing our data in the first place, and taking advantage of simple metadata tools that are already available to us. Few organizations have a standard system of nomenclature for naming digital data files and folders, and no industry standard for digital data file nomenclature exists. To be sure, this is an epidemic problem that extends far beyond the building industry, but it severely hampers our ability to organize and retrieve building information. In most organizations, individual employees or project teams name digital data files using whatever naming "system" makes sense to them. Any such system rarely extends across all data file types and is often haphazardly enforced. Most file naming systems fail to take into account that a file may be moved from its original file folder and that the file name—outside the folder—may not be sufficiently descriptive of the file contents. To extend our library analogy, our failure to maintain file naming conventions is equivalent to publishing books without meaningful titles.

Ideally, we would have an industry standard data format for file and folder naming similar to the data format standard for card catalogs used by libraries. But in the absence of an industry standard, individual firms can do a much better job of maintaining consistent file and folder naming nomenclature. The problem can be easily remedied with simple file and folder naming conventions.[1] By documenting their file and folder naming protocols, individual business enterprises can transfer that information—along with the files and folders—to third parties, who can then use it to understand how the information is organized.

Even fewer organizations take advantage of the data fields in the "File Properties" that are available for every electronic file type—the electronic equivalent of a card catalog record. The data fields in these file properties are a form of structured information at the metadata level; when used, they provide both a human- and machine-readable summary of the information contained in the digital file that can be read without opening the file. File properties can be used to provide summary information about the contents of the file but at a more detailed level than the file name. This information management feature is available to everyone—and has been available for years—at no additional cost and requires no additional software.

Information management software applications with sophisticated indexing and searching algorithms can help us overcome much of the mess of information we have created by helping us compile and retrieve relevant information from an undifferentiated mass, but there is a limit to how even the best information management or project management tools can manage undocumented and inconsistent business processes. Even such powerful tools cannot, on their own, completely solve fundamental business process and information exchange problems described in this chapter. As an industry, that is our collective responsibility.

Case Study: Information Management

One of the first projects assigned to one of the co-authors early in his career was to develop a drawing index for an organization that had over one million drawings on file. Some of the drawings still existed in their original form, on linen or vellum, but many had been converted to either 105mm or 35mm film. The drawing archive spanned multiple facilities that had inconsistent data collection and retention practices. The data for most facilities consisted of original construction documents as well as documents for multiple subsequent renovation and modernization projects. The "complete data set" for some buildings could span a hundred years. A search through the drawing archive to obtain drawings relevant to a new project could take weeks. The person conducting the search could not be certain that all relevant drawings had been found, and many drawings, once retrieved and examined, would prove to be irrelevant to the new project. Consequently, confidence in the quality of the information was very low.

The indexing project began by collecting a few pertinent pieces of information about each drawing: the facility name, number, and location; the contract number; the drawing number, and the type of drawing (e.g., architectural, structural, mechanical). The data was collected on 80-character punch cards, with the data for each drawing limited to 160 characters (spanning multiple cards by using the drawing number as a unique identifier).

After 10,000 drawings had been catalogued, the indexing team realized that the quality of some of the drawings was so poor that they were almost unusable. Recognizing that even poor-quality material could be of potential value, but not wanting every search to result in the printing of vast quantities of unusable material, the team decided to add an additional data field to each record indicating the quality level of the drawing. This enabled searchers to retrieve the highest-quality information first, and retrieve information of lesser quality only if it were needed or there was no other alternative.

Once completed, the indexing system was in place for many years until all the drawings were scanned and made available almost instantly in digital form. Over time, a non-value-added business process that initially required weeks to complete and yielded poor results was reduced to hours and then minutes as new technologies became available. However, the key to improving the business process organizing and cataloguing the information in a structured way is with metadata tags. Despite the emergence of new technologies, the information would have remained largely useless without the original indexing effort.

INFORMATION PROVENANCE

In order for the information contained in a BIM to be considered reliable, its provenance must be sufficiently well established to assure anyone using it that it faithfully represents the real-world conditions of the building itself, just as the authenticity of a great work of art is often established by its provenance—the evidentiary record of the work's chain of custody from the original artist to the present owner. Who the intermediate owners may have been in the life of a painting is important only to the extent that each owner represents a link in the unbroken chain. A break in the chain of custody breeds doubt about a work's authenticity. Doubt erodes the value of the entire data set of information. If a single link in the chain of custody of a great painting is missing, the other links have no value whatsoever, no matter how reputable the owners. The authenticity of the painting must be confirmed through other means: by laborious analysis of its age, style, materials, or painting technique. If the chain of custody is unbroken, such forensic analysis is unnecessary.

Information provenance is a vital component of information assurance. Few people will be willing to use information later in the life cycle of a building if its provenance is not known. As with the provenance of great artworks, exactly *who* the previous stewards of building information may have been is relatively unimportant. What matters most is whether the information has been continuously maintained by responsible stewards to reflect real-world conditions accurately.

FIGURE 6.2
**Palazzo Dugnani, Sala
del Tiepolo, Milan.**
Image courtesy of
Professor Ezio Arlati,
Politècnico Di Milano.

A vital component of every data set in an information exchange is its provenance, to assure anyone who has need of the information that appropriate care has been taken in compiling and verifying it (Figure 6.2).

Establishing the provenance of building information immediately raises the specter of liability for the original authors of that information, particularly for design professionals, for whom professional liability may extend far beyond the period of their responsible control of the information. Fearful of potential future liability, some design professionals will strip building information they have authored of any identifying information before conveying it to a third party, immediately rendering it worthless. The provenance of building information must be sufficiently clear and complete to permit anyone intending to use it to judge the reliability of it for themselves. The identity of each individual link in the chain of custody of building information is necessary to validate the integrity of the chain itself.

The building industry as a whole is clearly moving toward a new business environment—supported by new forms of legal agreements—in which the burden of responsibility for relying upon building information is shifting, under certain defined circumstances, from the original author to the person using it. In this environment, the liability of the original author for the use (or misuse) of building information terminates at the moment the original author no longer has responsible control of it. The original author retains liability for use of the information in its original, unaltered form for its originally intended purpose, but recipients are solely responsible for any other use or any subsequent modifications. Among the standard contract forms now available for establishing the terms of digital information exchange is AIA Contract Document E201™–2007, "Digital Data Protocol," designed to be attached as an exhibit to an agreement between two parties, and AIA Contract Document C106™–2007, "Digital Data Licensing Agreement," for digital data exchange between parties who are not otherwise in privity. More recently, the AGC-led ConsensusDOCS consortium has released the ConsensusDOCS 301 BIM Addendum, a document that addresses issues such as who owns the model, how information is managed, and how risk is allocated, while the AIA has released AIA Document E202–2008 BIM Protocol Exhibit, a scope definition document that defines the level of detail to which individual elements of the model will be developed and identifies the model element author for each element. In addition to providing the framework for new business relationships, all of these standard contract documents enable business partners to document information provenance properly.

INFORMATION MATURITY

The quality, detail, and reliability of individual pieces of information collected will vary widely throughout the life cycle of buildings, even if rigorous data collection and conservation protocols are implemented. People will always have a tendency to collect and compile only the information they need to do their jobs, and few tasks require detailed or complete building information. Still, any information gathered by authoritative sources can be valuable—even if it is of a general nature or incomplete—provided that its quality level is defined relative to other pieces of information. If you know who collected a particular piece of information (its provenance) and for what purpose, you can then determine how useful that information is for other purposes. The quality level is another piece of metadata.

Information generated at even the earliest stages of any business process should be tracked as it matures, producing a continuous record of when and how it has changed. Many software applications already have change management and tracking features that can help building industry professionals maintain a record of the evolution of information. Web-based data backup systems can automatically maintain a permanent record of every version of every data file. Users can further augment these tools with methodical data file archiving protocols at significant milestones.

The Wayback Machine: Archiving the Web

If the prospect of establishing data file archiving protocols seems daunting, visit the Internet Archive Wayback Machine at www.archive.org/web/web.php. This site archives over 85 billion Web pages from 1996 to several months ago, including every revision ever made to every Web page. You will be surprised at the ease with which you can find pages that no longer exist on your own Web site. Surely if someone has figured out how to archive virtually the entire World Wide Web and make it available to the world at no cost, it must be possible to develop robust data file archiving protocols for your business enterprise for a reasonable price.

A good example of a type of information that "matures" throughout the design and construction phase of buildings is construction cost estimating. A cost estimate may be completed several times on a typical design and construction project, with the cost estimate based on increasingly detailed and prescriptive information at every successive stage. In the early stages of building design, an architect or constructor may make an estimate of construction cost based on the average cost per square foot for buildings of comparable type and quality. While such an estimate represents little more than an educated guess based on many undocumented assumptions, it may well be sufficiently accurate for its intended purpose: to decide whether the project is sufficiently viable to warrant further development. If the level of information maturity is documented, the cost of the facility becomes a piece of the information that the owner can use at many stages of strategic decision making, rather than being a piece of information that is considered reliable only in the late stages of project development.

In the early planning stages of a project, such a rough estimate represents an appropriate allocation of professional resources. A more detailed, prescriptive estimate would be a waste of time and money. Among the important pieces of information about that estimate are the identities of the persons who generated it, their level of expertise and judgment, and the purpose for which the estimate was prepared. Architects, engineers, and constructors who understand their markets well and have a firm grasp of their clients' quality expectations can often predict actual construction costs with a high degree of accuracy on the basis of the sketchiest project information. Should the cost estimate they generate and control of the project pass from their hands, however, the estimate may be of no value at all, since another project team might have a completely different set of underlying (and equally undocumented) assumptions. But if enough is known about how, why, and by whom the cost estimate was prepared, then the estimate might still be of value to others (e.g., "wood-frame multi-family housing of up to three stories designed by architect 'A' and built by constructor 'C' in locale 'L' will cost 'D' dollars per square foot to build"). Comparative analysis of the same information by other estimating teams suddenly becomes very valuable.

At the highest level of information maturity, a final construction cost estimate is likely to be based on a prescriptive list of materials, equipment, and systems that is developed from a fully detailed and documented design, with very little of the cost based on the assumptions, industry averages, or rules of thumb used for the initial estimate. The process by which a cost estimate evolves from a "best guess" based on assumptions to a reliable estimate based on a detailed design can be documented and the "information maturity level" assigned to the estimate at each stage in its evolution. The information maturity level is itself a valuable piece of metadata that informs project participants of the degree to which they can rely on the information for decision making.

Another aspect of information maturity is its continuous evolution and development. For each construction cost estimate described above, the earlier, "immature" information is typically discarded, and the estimating process begins anew at every stage. Instead of being discarded, structured information should evolve through successive levels of maturity, so that use can be made of all previous information that might be useful. The documentary record of the evolution in information maturity can then be used to refine the workflow of future projects. In our cost estimate example, an analysis of the estimate-development process might prompt the estimating team to change its initial assumptions or alter the sequence of cost information gathering on a future project.

Information maturity levels can be assigned by original authors to any piece of information created in any stage of the building life cycle. A simple numerical system can be used by authors to assign an information maturity level to discrete pieces of information, which others can then use at later stages to assess the value of the information for another, unrelated purpose. In this way, both an initial cost estimate, which might be Information Maturity Level 1, can be preserved alongside the highly prescriptive, documented, and reliable information of the final cost estimate, which might be Information Maturity Level 10. The concept is part of the evolving National Building Information Modeling Standard (NBIMS), which will eventually provide building industry professionals with an objective information maturity scale that can be applied across all projects.

INFORMATION CONTENT DECAY

At some point in every business process, information reaches its most accurate or most mature stage, after which its maturity begins to decay. In a traditional design-bid-build project delivery process, for example, the construction documents issued for bid represent the highest level of information maturity about that building up to that stage and can be fully relied upon as an accurate representation of the as-yet-unbuilt building. Regrettably, this also is the moment of greatest information maturity in the life cycle of most buildings today. Immediately following contract award and as the project moves toward and through construction, the maturity level of the construction documents continually declines, and no other compilation of building information in the building's life cycle ever comes close to being as comprehensive or complete. During construction, modifications to the design will be made for countless reasons, as will substitutions in specified materials, equipment, and products. The documentation of these changes takes place entirely outside and apart from the methods used to prepare the original construction documents. By the time the building is constructed and occupied, the original construction documents can no longer be relied upon to represent real-world conditions accurately. The complete set of construction documents and subsequent modifications is typically a large body of scattered and conflicting information that can only be verified by field inspection, which rarely occurs. Many construction contracts require that the constructor deliver to the owner an as-built set of drawings, but the contract rarely specifies what is meant by "as-built record drawings," and in any event the drawings rarely encompass the full set of original contract documents. Even when a complete documentary record is transferred to facility managers, the mass of unstructured information is typically too unwieldy to be used efficiently. So facility managers typically pick and choose from the construction documents, recreating a portion of the design and construction data in a facility management software application before adding still more information of an operational nature. From this point forward and for the remainder of the life of the building, the body of building information typically remains fragmented and incomplete.

INFORMATION ELECTRONIC DEGRADATION

Nearly *all* building information created today is originally created in electronic form. No one in the building industry is sitting at a manual typewriter or a drafting board, at least not anyone who wishes to remain in business much longer. The type of information created electronically includes:

- Building geometry
- Manufacturers' product information, including warranty information
- Building systems information
- Building performance information
- Prescriptive or performance specifications
- Furnishings, fittings, and equipment information
- Construction cost information
- RFIs, change orders, other correspondence containing building information

Despite the fact that all of this information is created electronically, it is routinely degraded to nonelectronic form, sometimes immediately after creation, for one purpose and one purpose only: to be transmitted to another party. So our principal challenge, then, is not to automate our data creation processes, but to automate our data exchange processes. Simply recognizing that all building information is available in digital form is a huge leap forward, as it will lead us to demand that information be provided to us electronically. Chapter 7 describes just a few of the many industry-wide standards development efforts underway to address the problems of information decay and degradation.

INFORMATION INTEGRITY AND CONTINUITY

Building information modeling opens the door to maintaining an accurate and more complete documentary record of building information throughout the building design and construction process. At its current stage of development, designers and constructors using BIM in a collaborative business environment can overcome the significant information decay that occurs between the design and construction stages, provided that the database capabilities of BIM are fully utilized. This, in turn, increases the quality of information available for transfer to facility managers at the end of the construction stage. If the information decay that typically occurs at these two significant moments of transfer of information stewardship can be overcome, the building industry will be well on its way to achieving the vision of life cycle information management.

The continuity of information and workflow that BIM fosters provides a powerful incentive for early and intensive collaboration of designers and constructors, both of whom stand to benefit from the collaboration (see Figure 6.3).

FIGURE 6.3
Building Information
Continuity Illustrated.

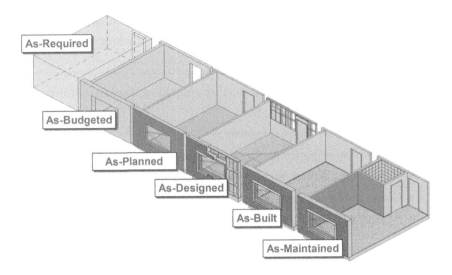

BIM enables architects and engineers to develop models to a much higher level of information maturity than was possible with CAD, which in turn reduces the number of changes that need to be made during the construction phase. This is especially true with respect to physical interferences, or clashes, between building elements. CAD drawings, like their paper predecessors, are necessarily diagrammatic, which can make physical interferences between building systems very difficult to detect. With BIM, the opposite is the case. If a building information model is developed to a sufficient level of detail, physical interferences are difficult to miss. As a result, problems that might become apparent only in the field with CAD can be discovered and resolved with BIM well before construction begins. This inherently reduces designers' potential liability for significant design errors.

An integrated model free of interferences is an obvious boon to constructors. In addition, BIM can be as useful to the constructor for building as it is to the architect and engineer for designing. BIM can significantly aid constructors in determining the cost, constructability, and construction sequence of the design, radically reducing cycle time throughout the construction process. Because building information modeling can support these core business processes of constructors in a way that CAD never could, constructors are incorporating BIM into their business processes at a very rapid pace. This increases the likelihood that a building information model—the repository of design information—will become constructors' preferred repository for construction information.

The intensive collaboration that BIM fosters makes it possible for design and construction teams to conceive of creating an "advance as-built model" of a building in BIM and then "building to the as-built;" the building is built virtually in BIM, and then again in reality. The timely availability and arrival on the job site of specified materials, products, and equipment remain wild cards that could still result in field modifications to the original design, but this too can be controlled through more intensive supply chain management.

INFORMATION TRANSPARENCY, ACCESSIBILITY, AND SECURITY

The issues of information transparency, accessibility, and security are interrelated. When building information is available to all business partners, its transparency can positively affect behavior. As Admiral Thad Allen, Commandant, United States Coast Guard, has often been quoted, "Transparency of data yields self correcting activity." Whenever business partners have access to the data of all other partners and can see the status of mission-critical information flows, it compels everyone to perform at the highest possible level without the need for alerts, reminders, or coercive management measures. The transparency of information flow in most project management software applications is an excellent example. When you realize that the completion of a task by another business partner is dependent on the completion of your own task, you have a powerful incentive to complete your task and maintain the flow of information.

Any structured compilation of information is inherently more valuable and its misuse potentially more dangerous than the same information in unstructured, decentralized form. For this reason, information security must be managed strategically and not left to chance. Access to information must be managed not just for security reasons, but also to ensure its integrity. The U.S. National BIM Standard, version 1, part 1, recommends that an information assurance manager, or BIM manager, be designated for every project. We would prefer to call this individual the *information steward*. This person would be responsible for maintaining the chain of custody of building information, allowing access to authoritative sources and restricting access of unauthorized persons through user IDs, passwords, and other access tools, and ultimately transferring the building information model to a successor information steward in the building life cycle in an official and legally documented form.

Flexible information management solutions are available today that allow information managers to decide which information will remain behind an organization's firewall, available only internally, and which information will be made available to business partners externally. These rule-based tools require relatively little active management by the information steward once the access protocols have been determined. Another tool in the information management toolkit are file viewers and view-only file formats, which allow the authoritative sources of information to edit the original data files and publish them in a view-only format, which allows other business partners to view, review, and comment. These simple tools, now widely available for many file types, allow communication and information exchange to take place between business partners electronically while preserving the integrity of the information.

Controlled access to building information by multiple business partners is no longer the challenge it once was. Increasingly, building industry software developers are deploying service-oriented (software) architecture (SOA), a method for systems development and integration where functionality is grouped around business processes and packaged as interoperable services.[2] SOA uses Web services standards and technologies and is rapidly becoming a standard approach for business enterprise information systems.[3] SOA enables companies to make their IT systems available to internal or external users, allowing applications to be used by different groups of people both inside and outside a company who have the requisite credentials for information access. It is far superior to the file-based client/server software architecture on which many software applications are based and which were developed without regard for a collaborative working environment.

While we tend to think of information security in terms of information integrity or threats to life safety, nearly every building owner has an interest in controlling access to certain pieces of building information for business reasons. The information security concern may be related to life safety threats, how a space is used, the location of vital utilities and utility service cutoff devices, or the locations of vital building systems, sensitive personnel or business activities, or valuable inventory. An effective and flexible information management system must take all of these considerations into account. Otherwise, information security concerns could severely hamper information flow, workflow, and improved efficiency.

INFORMATION FLOW

The desired attributes of building information described thus far in this chapter—provenance, maturity, integrity, continuity, transparency, and interoperability—are important not for their own sake, but to support greater efficiency and productivity throughout the life cycle of buildings through improved information flow. This can only be achieved through industry-wide cooperation to reform workflow throughout the building industry supply chain. Table 6.1 identifies just a few of the business processes in the design and

Table 6.1 Information Flow Processes.

Process	Current Practice	Desired Future Practice
Incorporation and integration of geospatial information (site data) with building information	Information is manually gathered from multiple sources, independently validated, and manually entered into a CAD format.	The building information model is geospatially located; digital geospatial information obtained from authoritative sources and integrated with the model as needed. Metadata is used to ensure validity and accuracy. During construction, the GPS coordinates of site utilities and other site features are captured and incorporated into the model.
Retrieving and incorporating information from as-built record documents	As-built record drawings are regarded as unreliable. Existing conditions data is extensively recollected and validated or recreated.	As-built record drawings are regarded as accurate and reliable. At the start of a new project, as-built information is routinely audited to confirm accuracy then incorporated into the model for a new project.
Procurement of manufactured products	Manufactured products are manually identified and tallied from drawings and specifications. Product data sheets are compiled and approved by the designer of the specified system.	Specifiers property sets for manufactured products are incorporated directly into the building information model. A quantity take-off of the model generates a complete list of manufactured products. A matrix is generated that automatically compares the products of different manufacturers with the required specifiers property sets for each product. Requests for pricing and availability, selection and acceptance of vendor offers, and product orders are managed electronically.
Installation of manufactured products and equipment.	Proper installation is dependent on the knowledge and skill of the installer, which may or may not conform to the manufacturer's installation requirements. Improper installation, which may not be discovered for many years, voids warranty.	The data set for each manufactured product or piece of equipment includes detailed installation procedures and required certification of installers. Equipment is installed in accordance with manufacturers' requirements. Manufacturers' representatives are automatically notified of completed installations and verify conformance to manufacturers' installation requirements prior to acceptance by the general contractor or owner. Manufacturers' acceptance and warranty is incorporated into the model.
Procurement of fabricated assemblies	Few fabricated assemblies arrive at the job site; most construction is performed on-site.	Most raw materials and manufactured products are prefabricated into assemblies prior to delivery to the job site. On-site labor, construction time, and waste are significantly reduced.
Procurement of raw materials	Quantities of raw materials and often-undocumented accessories are manually tallied from drawings and specifications. Materials are ordered as needed in excess quantities to ensure adequate quantities on-site. Excess materials enter the waste or recycling stream.	A quantity takeoff of the building information model generates an accurate quantity of raw materials, including accessory materials, which are ordered in more precise quantities. Greater quantities of raw materials are incorporated into prefabricated assemblies. Fewer raw materials are delivered to the job site. Space needed for on-site storage of materials and waste is significantly reduced.
Tracking raw materials and manufactured products	Raw materials and manufactured products are manually tracked and stored in lay down areas on-site. Items are frequently lost, misplaced, damaged, or stolen.	Raw materials and manufactured products are tracked electronically in a centralized database from manufacture to installation. Real-time tracking supports just-in-time delivery, eliminating multiple handling, the need for lay down space, and shrinkage due to loss, damage, or theft.

construction information supply chain—which is itself only a small fraction of the building life cycle supply chain—that could be radically reformed and streamlined with better building information.

 ## THE LIFE CYCLE OF INFORMATION

Extending our view of building information from design and construction to the entire building life cycle reveals the full scope of information that needs to be gathered throughout the building life cycle. Table 6.2, loosely based on Table 31 of the OmniClass Construction Classification System (OCCS) developed by the Construction Specifications Institute,[4] is a summary of information that is typically collected or needed at each stage of the building life cycle.

 ## STAKEHOLDER VIEWS

Every stakeholder in the building life cycle has a need to view building information from his or her own business perspective, which means that each stakeholder represents an interface requirement. This is one of the most challenging obstacles to sharing building information, since so few information interfaces exist. Table 6.3 illustrates the widely divergent information needs of various stakeholders.

 ## INTEROPERABILITY

Operability is the fundamental characteristic of simple tools used to complete simple tasks very well. A pencil is operable. A screwdriver is operable. A software application is operable. *Interoperability* is the fundamental characteristic of tools that are designed to work together as part of an integrated system to complete complex tasks. Interoperable tools may be no more sophisticated in their design than merely operable tools. Their sophistication results from their ability to function as part of a system or sequence of tasks. In manufacturing, for example, thousands of automated tools exist that can perform simple tasks such as measuring, cutting, pouring, grinding, crushing, drilling, polishing, rolling, clamping, screwing, stamping, heating, cooling, painting, curing, or drying. Each tool is designed to do one thing very well. These simple tools become interoperable when they are designed to work as part of an integrated, automated system in a defined sequence. Behind nearly every advancement of

Table 6.2 The Life Cycle of Information.

Life Cycle Phase	Information Available, Needed, and (Ideally) Compiled
Project Conception and Planning	Geospatial information related to terrain; vehicular access and traffic volume; demographics; utility connections; existing conditions; subsurface conditions; weather and climate including sun angles and prevailing winds; seismic conditions; and flood zone data.
	Economic data may include labor availability, cost of financing, and material availability.
	Other site information would include building information (and, ideally, building information models) about any existing buildings on site.
Building Design	Geometry of the building to determine length, height, and width, of structural, mechanical, electrical systems; building materials information including design and performance specifications, cost, and sustainability information; access and egress information, energy requirements, and other code related information.
Procurement and Construction	Detailed information about installed equipment including model number, serial number, installation requirements, certifications required for installation, warranty, and operating requirements; planned schedule of delivery and other construction sequencing information; cost of installation and equipment required for installation.
Building Commissioning	Design calculations indicating performance requirements of all equipment. This phase is an opportunity to ensure that the information contained in the BIM is accurate and correctly depicts the true as-built condition of the facility, indicating where all items are installed.
Building Occupancy	Information for way-finding during normal occupancy as well as during emergency scenarios, including shelter-in-place. Information about utilities including power and communications to support moves, additions, and changes.
Facility Management	Preventative maintenance requirements; records of all maintenance required, including lists of spare parts needed for routine maintenance and to maintain operational status of all equipment; operating conditions including the specifications provided during commissioning to ensure operating ranges are sustained.
Facility Retirement or Reuse	All of the information about the facility that has been compiled in the Building Information Model, which includes geospatial information, design, construction, commissioning, occupancy, operations, and maintenance.
Life Cycle Closure	All of the information about the facility that has been compiled in the Building Information Model, which includes geospatial information, design, construction, commissioning, occupancy, operations, and maintenance, especially information needed for strategic planning of material recycling and disposal.

high technology lies just such a sophisticated sequence of low-technology tools. A highly automated assembly line is nothing more than an elaborate sequence of very simple robotic tools. Even the most advanced software is built upon binary code; every seemingly sophisticated function is a product of an elaborate sequence of yes/no, on/off binary operations. Interestingly, this aspect of high technology mimics the sophistication of biological ecosystems discussed in the previous chapter. Each machine in a high-tech system or binary operation in a high-tech software application "accepts" the output of the previous

Table 6.3 Stakeholder Views.

Stakeholder	Information Needs
Owners	High-level summary information about their facilities
Planners	Existing information about physical site(s) and corporate program needs
Realtors	Information about a site or facility to support purchase or sale
Appraisers	Information about the facility to support valuation
Mortgage Bankers	Economic data, demographic data, information about financial viability of the facility
Designers	Planning and site information
Engineers	Electronic building information to import into engineering design and analysis software
Cost and Quantity Estimators	Electronic building information to obtain accurate quantities
Specifiers	Electronic building information, including intelligent objects, to develop specifications and link to later phases
Contracts and Lawyers	Accurate legal descriptions
Construction Contractors	Electronic building information, including intelligent objects, for quantity takeoffs of materials and equipment, construction cost estimating, subcontractor and material bidding and ordering; a repository for construction information
Subcontractors	Same as for contractors, including clear communication of subcontract requirements
Fabricators	Electronic building information, including intelligent objects, for numerical controls for fabrication
Code Officials	Electronic building information for automated code checking
Facility Managers	Building product, materials, and equipment warranty and maintenance information
Maintenance and Sustainment	Electronic building information to identify easily products needed for maintenance and parts for repair or replacement
Renovation and Restoration	Accurate electronic building information to minimize or eliminate unforeseen conditions and the resulting unanticipated cost
Disposal and Recycling	Detailed building information about installed materials, including information about what is recyclable
Scoping, Testing, Simulation	Sufficiently detailed electronic building information for accurate simulation of real-world conditions
Safety and Occupational Health	Sufficiently detailed electronic information about installed materials to prepare Material Safety Data Sheets (MSDS)
Environmental and NEPA	Sufficiently detailed electronic building information to conduct environmental impact analysis
Plant Operations	Sufficiently detailed building information model for 3-D visualization of plant processes
Energy, LEED	Sufficiently detailed electronic building information for energy analysis and design optimization
Space & Security	Sufficiently detailed building information model, including intelligent objects, to analyze and guard against vulnerabilities
Network Managers	Sufficiently detailed 3-D building information model for IT network design and troubleshooting
CIOs	Sufficiently detailed electronic building information to enable sound decision making regarding IT infrastructure
Communications	Sufficiently detailed electronic building information to enable support for moves, additions, or changes.
Risk Managers	Sufficiently detailed electronic building information to assess and minimize potential risks
Occupant Support	Sufficiently detailed 3-D building information model for wayfinding
First Responders	Sufficiently detailed 3-D building information model for emergency wayfinding to minimize loss of life and property

machine or operation as a "found object" or "natural resource." The individual tools "communicate" with one another entirely through their environment. One tool deposits its work product in the exact location where the next tool is designed to pick it up; the output of one software operation is the input for the next one.

A pocketknife is an excellent example of a simple tool designed to do one thing very well: make small cuts in relatively soft, everyday things such as rope, canvas, fishing line, or even fish. A Swiss Army knife, only slightly bulkier than a simple pocketknife, is designed to do many simple things well. In addition to a knife blade, it may include a corkscrew, a screwdriver, a bottle opener, a magnifying glass, a pair of scissors, an awl, and even a pair of tweezers. But a Swiss Army knife is not an interoperable tool; it is a bundle of individual, unrelated tools, designed for a variety of unrelated tasks. As more and more features are added to a Swiss Army knife, its most salient characteristic as a pocketknife—its compact size—is eroded and each individual tool becomes more difficult to use. You may have to open three or four of the tools—in the woods, at night, in the rain—before you find the one you need.

Most software applications available to the building industry today are operable but not fully interoperable. While there are notable exceptions, very few software applications are designed to receive information routinely and reliably from another software application or transmit information to a third. Most building industry software is designed on the presumption that it exists in its own little world. Whereas technology in other industries has moved consistently toward greater interoperability, technology in the building industry has generally moved toward the bundling of a greater and greater number of tasks—what we commonly call "feature creep." As tools become more complex they become slower to operate, more difficult to learn and use, and less useful for their originally intended purpose.

Of the building information exchange challenges described in this chapter, interoperability is the only challenge that is primarily the responsibility of software developers to fix. But in all fairness, most software companies allocate their software development budgets in response to customer demand, and as customers, building industry professionals have demanded more and more features, not interoperability. As BIM technology develops there are hopeful signs that we are moving away from this unproductive path of technology development, but as an industry we are still a long way from completely shaking off the Swiss Army Knife syndrome[5] (see Figure 6.4).

Interoperability can be achieved in any number of ways. Software developers can agree to embed support in their software applications for open-standard data formats, such as the Industry Foundation Classes (IFCs) of building

FIGURE 6.4
Feature Creep: The Swiss Army Knife Syndrome.
(Photo by Berteun)

SMART International, which allows a defined data set of building information to be exchanged between software applications with dissimilar proprietary data formats.

Another open standard created specifically for the steel industry is the CIMSteel Integration Standards (CIS/2), a product model and electronic data exchange file format for structural steel project information.[6] CIS/2 is the work of the Eureka CIMsteel (Computer Integrated Manufacturing for Construction Steelwork) Project, a European steel industry coalition.[7] The CIS/2 format captures all the information related to the design, analysis, procurement, fabrication planning, fabrication automation, and erection of structural steel in buildings.[8] In 1998, the American Institute of Steel Construction (AISC) endorsed CIS/2 as the preferred format for data exchange between steel-related software, signed an agreement with the developers of CIS/2, and embarked on an aggressive effort to convince software developers to support the data-exchange standard. As a result, many software applications for steel design, analysis, engineering, fabrication, and construction now include CIS/2 file import/export capabilities. The complete list of software applications that support CIS/2 can be found at the Georgia Tech CIS/2 Web site.[9] The story of CIS/2 demonstrates that when an industry segment bands together to express its need for interoperability clearly, interoperability soon becomes reality.

The IFC (ISO 16739), ISO 15926, and CIS/2 formats share common origins—all were developed using the International Standards Organization's (ISO) Standard for the Exchange of Product Model Data (STEP), which was developed to provide a mechanism capable of describing product data throughout the life cycle of a product, independent of any particular system.[10] Several STEP resource definitions are common to IFC and CIS/2, and both use the same STEP-related modeling language, EXPRESS, for developing and defining the model.[11] To achieve greater compatibility, AISC and buildingSMART International embarked on a harmonization project that resulted in a modification to the IFC model in IFC release 2x3.

Interoperability can also be achieved through eXtensible Markup Language (XML), an open-standard of the World Wide Web Consortium (W3C) to facilitate the structured exchange of information. XML can be used to support proprietary (nonpublic) or open-standard information exchanges. Chapter 7 describes one such open standard, agcXML, that is being developed by the Associated General Contractors of America (AGC) in cooperation with the buildingSMART alliance.

Finally, software companies can enter into agreements with one another to establish proprietary methods for information exchange, either through Application Programming Interfaces (APIs) or proprietary data exchange formats. While open, nonproprietary data formats offer the greatest potential for interoperability among many different software applications, any data exchange format that fosters reliable interoperability benefits the industry by providing a channel for electronic information exchange.

ENDNOTES

1. Lachmi Khemlani, "The Cis/2 Format: Another Aec Interoperability Standard," AECbytes, http://www.aecbytes.com/buildingthefuture/2005/CIS2format.html.

2. Ibid.

3. Ibid.

4. OCCS Development Committee Secretariat, "Omniclass: A Strategy for Classifying the Built Environment," Construction Specifications Institute, http://www.omniclass.org/index.asp.

5. "Swiss Army Knife Wenger Opened," Wikimedia Commons, http://commons.wikimedia.org/wiki/Image:Swiss_Army_Knife_Wenger_Opened_20050627.jpg.

6. National Institute of Standards and Technology, "Cis/2 and Ifc—Product Data Standards for Structural Steel," National Institute of Standards and Technology, http://cic.nist.gov/vrml/cis2.html.

7. Khemlani, "The Cis/2 Format: Another Aec Interoperability Standard."

8. Ibid.

9. Charles Eastman et al., "Cis/2 Exchange Capabilities: Translator Interchangeability Matrix," Georgia Institute of Technology, http://www.coa.gatech.edu/˜aisc/index.php?cat1=1.

10. Wikipedia, "Iso 10303," Wikipedia, http://en.wikipedia.org/wiki/ISO_10303.

11. Khemlani, "The Cis/2 Format: Another Aec Interoperability Standard."

CHAPTER 7

Building Information Exchange Requirements

Our dilemma is that we hate change and love it at the same time; what we really want is for things to remain the same but get better.

—Sydney J. Harris

The preceding chapter outlines the issues and challenges relating to the quality of electronic building information—how it is organized and what its attributes need to be. This chapter is devoted to the content of information exchanges. Business processes in the building industry have developed over many years, and though many processes and workflows are often regarded as "standard" practices or procedures, very few of our business processes have actually been documented. Though the workflow and content of information exchanges may be consistent enough to be recognizable as distinct business processes, individual practices vary considerably from one organization or group of business partners to another. Variations in practice may represent only a small fraction of any individual business process, but they are significant enough to stymie automation of workflow and information sharing across the industry.

 THE BIG PICTURE

The industry is working to demonstrate how the day-to-day exchanges of information between practitioners can be documented. This chapter provides several examples of how this is being accomplished. What is currently missing is the overall view of the industry showing all the information exchanges that will ultimately need to be documented. Every activity in the industry that is ultimately required to produce a facility needs to be identified. Every exchange of information between those activities needs to be documented. This high-level task needs to be accomplished for two reasons: so that we know just how big a job we have in front of us, and so that we will know when we are done. This is certainly not a job for any one organization, but something that the industry must work together to produce. Once the "big picture" is defined we will have a much better idea of how the business process models for individual components of the building life cycle—such as those developed by the Charles Pankow Foundation and the Pre-cast Concrete Institute for precast concrete (described in Chapter 4)—will fit into the overall business process of the entire industry. In terms of our overall "business process maturity," we are currently at the stage of honing individual tools and processes. As we assemble these into an overall picture, we will be able to manage the transformation of the entire life cycle and finally determine what the total cost and benefit of that transformation will be.

INFORMATION DELIVERY MANUALS

If our business processes are to be automated, the required content of information exchanges must be defined. While the Industry Foundation Classes (IFCs) of buildingSMART International provide a common exchange standard for electronic building information, the IFC data format alone is insufficient, because it does not describe individual business processes within the building life cycle or the information needed to complete them.

In recognition of this gap, Statsbygg, the government agency that manages the real estate portfolio of the government of Norway (the Norwegian counterpart to the U.S. General Services Administration) launched a buildingSMART project to develop an Information Delivery Manual (IDM) in 2005.[1] For industry professionals, IDMs will provide a plain-language description of business processes, the information requirements for each process to be carried out successfully, a description of additional information that the person executing the process may need to provide, and the expected outcome of each process. For software developers, IDMs will identify and describe the detailed functional breakdown of each process and the information-exchange requirements for each process.[2]

Figure 7.1 shows the role of IDMs in the information exchange process. End users define the content of information exchanges in plain English. This provides software companies with a clear definition of the information exchange requirements that their software applications need to support. Software developers then use an electronic information exchange mechanism—preferably, an open-standard data format such as the IFCs—to support the reliable and accurate exchange of information between software applications. Because the data set is defined and documented, end users know exactly what information is included in the information exchange.

Figure 7.2 depicts a simple information exchange in the early stages of the building life cycle, in this case the information exchange that takes place between a contractor and a supplier to purchase and install doors. While the transaction is simple, all three components of information exchange depicted in Figure 7.1 are needed to execute the transaction electronically and capture the information generated by the transaction for the remainder of the building life cycle. The transaction begins when the contractor extracts specification information (i.e., size, material, quantity, or fire rating) contained in the building information model to place an order with the supplier or manufacturer. (Note that if the actual order process is proceeded by a request for pricing process, yet another workflow process and set of information exchange requirements would have to be defined.)

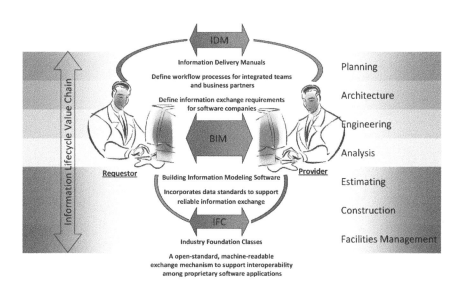

FIGURE 7.1
The Three Components of Information Exchange.
(Source: AEC Infosystems and ONUMA Inc. Used by permission.)

Information conforms to programmatic and project requirements or constraints, such as role of item in obtaining desired level of LEED certification.

Information is structured according to industry standards (NCS, IFCs, OmniClass); business partners do not need to negotiate either content or electronic data format.

Information
exchange:
purchase and
receive doors.

INPUT: information
contained in the
building information
model (e.g., door size,
material, quantity, fire
rating).

Control

Input

BIM

BIM
Output

The product of the information exchange (e.g., actual item and quantity delivered; cost; installation and maintenance instructions; warranty information) is incorporated back into the building information model in a continuous spiral of information gathering.

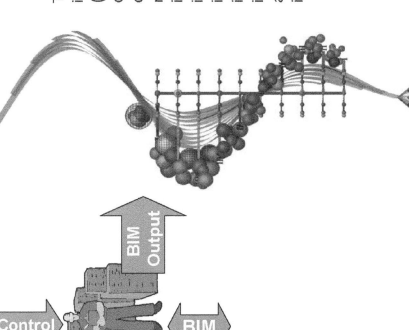

ONUMAINC.

FIGURE 7.2 Simple Information Exchange during Construction. (Source: AEC Infosystems and ONUMA Inc. Used by permission.)

These prescriptive specifications, in turn, may be based on a set of pro-grammatic or project requirements and constraints, such as the desired level of LEED certification, and must be formatted consistently to ensure the reliability of the information exchange. Together, these define the boundaries, limits, or controls for the information to be exchanged.

The delivery of the doors is typically accompanied by a packing list that details the actual items delivered and their cost, and other information pro-vided by the manufacturer such as handling and installation instructions, oper-ating instructions, and warranties. Ideally, this information is captured at the time of the delivery transaction and incorporated into the model, which is con-tinually enriched by the output of individual business processes.

At the other end of the building life cycle, Figure 7.3 depicts an informa-tion exchange that occurs during building occupancy. First responders to life-threatening emergencies such as fire need vital information about such things as the quantity and location of hazardous materials that may be stored in the facility, or the occupancy and likely location of handicapped persons. In this information exchange, a reliable, accurate building information model contain-ing information organized in accordance with recognized standards is made available to the first responders at the critical moment of response, but access to the information by the first responders expires once the emergency has passed to ensure that the information is not disseminated for any other purpose.

Simple workflow diagrams such as these can be developed for every busi-ness process in the life cycle of a building. In current practice, these activities and the content of the related information exchanges are continually and often poorly redefined. Small variations in practice become constant sources of mis-communication and enormous impediments to increased efficiency, productiv-ity, and effectiveness. IDMs are intended to define "best practice" business processes and the body of information resulting from those processes that should be stored in a building information model.

Each piece of software that a building industry professional deploys has a relationship to a certain set of activities in a business process and has embedded within it associated information flows. The information that the software needs to perform its function as well as the information that the software generates is of value in supporting that group of activities. The problem is that in a merely operable world there is no relationship between the input/output of one soft-ware application with other software tools.

This level of functionality, while it produces useful results, is also extremely wasteful since it causes information to be manually reentered into one software application after another. The non-value-added effort is significant; most of the inefficiency and waste in the building industry today can be attributed directly

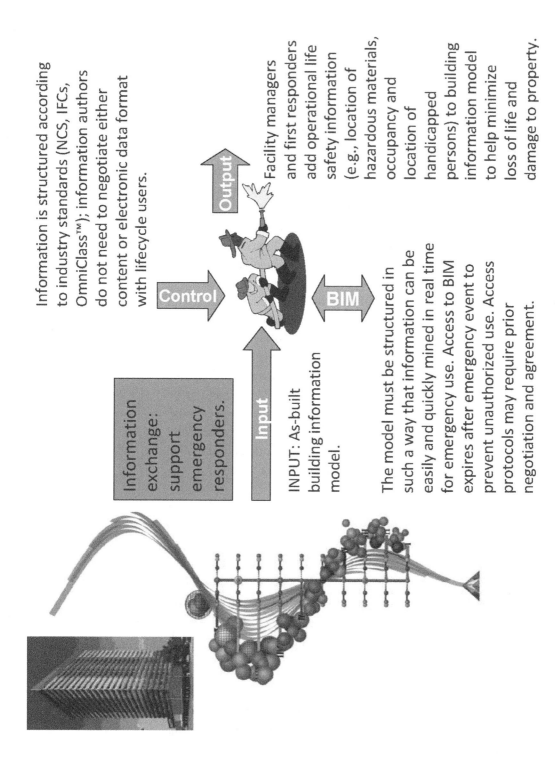

Information is structured according to industry standards (NCS, IFCs, OmniClass™); information authors do not need to negotiate either content or electronic data format with lifecycle users.

Control

Output

Facility managers and first responders add operational life safety information (e.g., location of hazardous materials, occupancy and location of handicapped persons) to building information model to help minimize loss of life and damage to property.

Information exchange: support emergency responders.

Input

BIM

INPUT: As-built building information model.

The model must be structured in such a way that information can be easily and quickly mined in real time for emergency use. Access to BIM expires after emergency event to prevent unauthorized use. Access protocols may require prior negotiation and agreement.

FIGURE 7.3 Simple Information Exchange for Emergency Responders. (Source: AEC Infosystems and ONUMA Inc. Used by permission.)

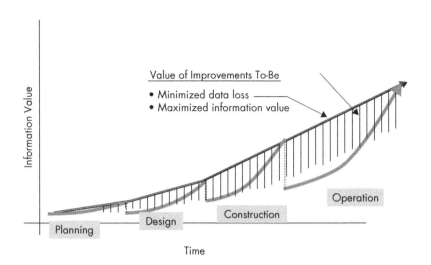

FIGURE 7.4
Data Losses in the Building Life Cycle.

to it. It often leads to information not being entered into "downstream" software applications—a symptom of both information content decay and electronic information degradation—such that subsequent business processes are literally starved for accurate and complete information, eroding the efficiency of those processes before they even begin. This has become an accepted norm in our operable environment, because it is simply too costly to do anything else. Hence, rules of thumb and other shortcuts are used to simulate or summarize information instead of exhaustively regathering complete and accurate information. Figure 7.4 graphically illustrates the information losses that occur throughout the building life cycle. The continuous line represents the "to be" condition in which all data is preserved for the next phase of the life cycle, for which only additional information needs to be added.

DEFINING "BEST CASE" BUSINESS PROCESSES

An inventory of the software applications used in any organization instantly reveals that no single software application supports all of the business processes and activities of the organization. The most interoperable set of applications is the "office suite," which typically consists of some combination of word processing, spreadsheet, visual presentation, e-mail, relational database, and desktop publishing applications. The interoperability of these applications is relatively high, in part because each software company in this market sector typically offers a complete office suite, but mostly because there is a great deal of consistency in data format among these applications, even between the software products of different companies. Office suites also support a wide range of graphic data formats. We take it for granted that we can import and export both alphanumeric and graphic information from one office suite application to another. This is the type of working environment that should exist across the building industry, though it would be unreasonable to expect that any one software provider will ever offer a complete "building industry suite" of applications. This heightens the need for nonproprietary data exchange mechanisms.

If a software application has a good "awareness" of how information will come to it, then it can minimize the effort needed to import that information. As the Information Flow and Information Life Cycle tables of Chapter 6 amply demonstrate, there is no business process in the building life cycle for which this critical need does not exist. Though information flows within individual business processes are challenging enough, when the flow of information from one business process to another in the building life cycle is considered, the challenge extends beyond the ability of software applications to talk to one another. Every group of stakeholders has its own jargon and views information from a different perspective. The units of measurement alone may be an obstacle to accurate and reliable information exchange. One stakeholder may measure building area in square feet while another thinks in terms of the number of beds. An architect or engineer may think of a concrete slab in terms of its length, width, and thickness, and how it connects or relates to other permanent structural elements. A construction cost estimator thinks in terms of cubic yards of concrete, how much temporary formwork is needed to support it, and how the formwork can be erected and dismantled.

The underlying data is the same; it is how each stakeholder views the information that is different. Software applications designed to support the business processes of each stakeholder cannot rely on the equivalent of "cut and paste" for information exchange. Building industry organizations worldwide are engaged in defining "best-case" business processes that will allow information to flow—reliably, seamlessly, and accurately—across these boundaries. It is a long, costly, and arduous task. A great deal of "data mapping" is required to translate, for example, square feet into bed capacity. Algorithms must be written for this purpose, and users need some assurance that the algorithms have been written to "translate" the data correctly. Simple mathematical calculations may suffice in some cases, while an intermediary exercise of professional judgment may be required in others.

A simple, perhaps familiar example that illustrates the complexity of the challenge is the business card scanner and its accompanying software. These hardware/software solutions are intended to automate a very simple task: enter the information on a printed business card into the correct data fields in a contact management software application. The accuracy and reliability of some of these tools, however, is remarkably high when you consider the lack of consistency in business card design. The challenge of reliable information capture is not the high-tech optical character recognition (OCR) scanning technology, which typically can interpret legibly printed text with a high degree of accuracy. The problem is that there is no standard for how information should be displayed, labeled, or organized on a business card. Is the area code of telephone numbers enclosed by parentheses? Are the ten digits of telephone numbers formatted with hyphens, periods, or spaces? What do the letters "m," "t," "f," "d," or "c" adjacent to a telephone number indicate? How do those letters "map" to the words "Main," "Mobile," "Tel," "Fax," "Direct," "Cell," or "Company"? Do these markers appear before or after the number? Does the company name appear as part of a logo, as text, or both?

A few simple "information format" standards for business cards would allow graphic designers continued creative freedom while vastly simplifying the algorithms needed to "interpret" the text on business cards and vastly increasing both the speed of the scanning process and the accuracy of the scanning results. The fundamental "information exchange requirement" for business cards is that information contained on the card be sufficiently legible and accessible so that business partners can contact you. If your business objective is to increase the likelihood that your contact information will become embedded in the information systems of other companies, then you might want to try this simple experiment: run your business card through one or more of the popular scanners and see if the information is correctly interpreted. In the building industry, the technology we have at our fingertips is not the problem. The core problem is our failure to organize our information.

AGCXML: ORGANIZING TRANSACTIONAL INFORMATION

A good example of an industry effort to organize information better is the agcXML Project, a buildingSMART initiative funded and led by the Associated General Contractors of America and executed by the National Institute of Building Sciences.[3] The outcome of the agcXML Project is a publicly available structured format for the information that is now exchanged during the design and construction process through any number of "standard" documents, beginning with an Owner/Contractor Agreement, extending through a Certificate of Occupancy, and including such critically important documents such as Requests for Information, Change Orders, and Applications for Payment.

FIGURE 7.5
Code Compliance
Checking Using SMART-
codes.
(Source: International
Code Council and the
U.S. Coast Guard. Used
by permission.)

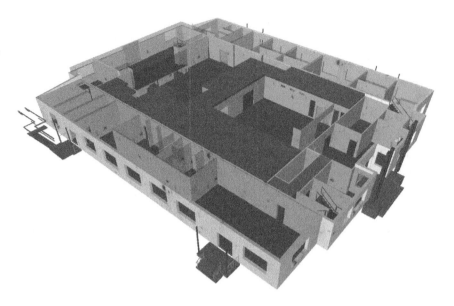

All of the information contained in these documents is now entered electroni-cally—and multiple times—into any number of software applications including contract document software, project management software, construction cost estimating software, and project accounting software. Some, but not all, of this transactional information is building information, but unless it is structured in a consistent, generally accepted format, it cannot be exchanged between busi-ness partners and software applications. The agcXML Project will result in a set of eXtensible Markup Language (XML) schemas that any software pro-vider can support relatively easily without having to disclose any proprietary software code. It will help close a significant gap in electronic information degradation.

SMARTCODES: AUTOMATING THE REGULATORY PROCESS

The International Code Council (ICC) is leading and funding a building SMART initiative to develop SMARTcodes, a tool to automate code compli-ance checking and identify items in and aspects of a building design that do not conform to established codes, standards, and regulations[4] (see Figure 7.5). ICC will collaborate with BIM software developers so that their applications can support the delivery of information that is sufficiently detailed and comprehensive to enable design professionals and code officials to use SMARTcodes for code compliance checking on any project, which is estimated will cut the time needed to review construction documents, issue building permits, and approve buildings for occupancy by 50 percent. While the benefits that will accrue to regulatory authorities can be measured directly, the greater benefit will accrue to design professionals, who will devote far less time to iterative revisions and submissions, and building owners, who will be able to reduce dramatically the cycle time and increase the schedule predictability of preconstruction activities. Once implemented, it will close yet one more yawning gap of information content decay and electronic information degradation.

The challenges described thus far, as great as they are, relate largely to attributes of information that can be parsed or quantified. Yet another layer of information management in building life cycle processes is the qualitative information found, for example, in construction specifications. How can installation requirements or performance requirements be associated with the object-oriented information contained in a building information model? The business processes related to qualitative information—specifications, referenced standards, warranties, operational requirements—need to move toward an object-based software environment that provides the building industry professionals who create and use that information with a familiar, easy-to-use interface. The problems grow as the building life cycle lengthens, because the body of information, the number of sources, and the number of stakeholders continually grows, as do the number of "downstream users," many of whose information needs drift further and further from what we would consider to be "building industry" business processes.

THE CONSTRUCTION OPERATIONS BUILDING INFORMATION EXCHANGE

An important industry effort toward capturing and structuring building information for use in building operations is the Construction Operations Building Information Exchange (COBIE), a buildingSMART alliance project initially funded by National Aeronautics and Space Administration (NASA) and the U.S. Army Corps of Engineers (USACE). NASA and the White House Office of Science and Technology Policy provided the first two grants that started this project in 2005.[5]

The Facility Maintenance and Operations Committee (FMOC) of the National Institute of Building Sciences (NIBS), chaired by Bill Brodt, currently Experimental Facilities Engineer for NASA's Facilities Engineering and Real

Property Division, formed a project team representing designers, builders, owners, commissioning agents, and software firms to identify the requirements for the information exchanges at the critical moment of construction-to-operations information transfer. The result—COBIE—is a data specification for information transfer. While the Industry Foundation Classes (IFCs) of buildingSMART International define the actual data elements of a building information model, COBIE is an excellent example of the "information management structure" that must be included with a building information model so that information can be compiled and exchanged most effectively and efficiently throughout the building life cycle at a high level of information maturity and with minimal information decay.

Brodt had long been involved with the SPECSINTACT or "Specifications-kept-intact" system that NASA and the U.S. Department of Defense (DoD) use for specification writing. A submittal register can be generated automatically from a SPECSINTACT specification, identifying all items that must be submitted for approval during the construction phase. This was a significant innovation in its own right that eliminated the time-consuming task of reading through a set of specifications to identify all required submittals and then typing them into a list.

Though the SPECSINTACT submittal register became the starting point for COBIE, it was designed so that a COBIE database could be initiated using any generic submittal register. The initial focus of COBIE was on tracking those items incorporated into a building that were under warranty. This information, which is rarely compiled in a methodical way, is vital to the management of a facility and has significant tangible value. If warranties are not properly tracked, a building owner could incur significant and unnecessary costs for equipment repair or replacement during the warranty period. By identifying the most critical instance of information decay and focusing on capturing the piece of information with the greatest potential return on investment, the COBIE project followed a familiar pattern of successful standards development:

- Define a clear business case for those stakeholders with an acute interest in solving the problem and who can provide the initial funding.
- Demonstrate the value of the standards-development effort through pilot projects.
- Expand the scope of the initial pilot effort to include a broader scope of data.

COBIE is the first project to establish structured format for compiling the information embodied in construction specifications with detailed warranty information subsequently provided by equipment manufacturers, spanning an important information-exchange gap. The COBIE Workflow Model illustrated in Figure 7.6 demonstrates that the flow of information through all phases of the building life cycle is critical to realizing the potential return on investment (ROI) of building information modeling.

The scope of COBIE quickly grew to include all of the information that is typically required to be conveyed to building owners at the end of construction—typically in unstructured form, as paper documents—including equipment lists, product data sheets, warranties, spare part lists, preventative maintenance schedules, and any other information needed to support the operations, maintenance, and management of the facility asset by the owner or property manager.[6] After several pilot projects at the U.S. Department of State Overseas Buildings Operations (OBO) office and the U.S. Army Corps of Engineers Seattle District Office, a COBIE specification was published to replace paper handover documents with structured COBIE data. It is currently required in a number of federal government design and construction contracts.

COBIE provides a mechanism for each authoritative source to enter data into a single database, simplifying the capture of this vital data. Design professionals enter design information, such as floor, space, and equipment layouts. Contractors enter the make, model, and serial number of installed equipment.

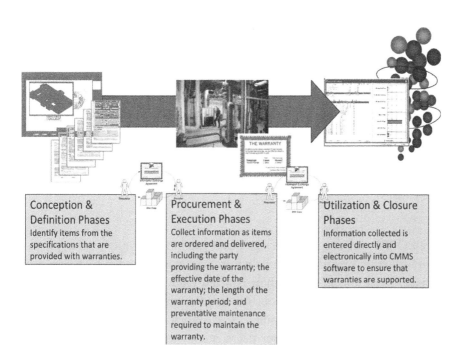

FIGURE 7.6
COBIE Workflow Model. (Source: AEC Infosystems and ONUMA Inc. Used by permission.)

Conception & Definition Phases
Identify items from the specifications that are provided with warranties.

Procurement & Execution Phases
Collect information as items are ordered and delivered, including the party providing the warranty; the effective date of the warranty; the length of the warranty period; and preventative maintenance required to maintain the warranty.

Utilization & Closure Phases
Information collected is entered directly and electronically into CMMS software to ensure that warranties are supported.

The core concept is to collect information needed for building operations and maintenance as it is created during design and construction. COBIE establishes the first workflow mechanism for methodically compiling information during one phase of the building life cycle—design and construction—and conveying it to users in another—operations—at a very high level of information maturity and reliability. It directly addresses the moment of greatest "information decay" in the building life cycle: the one that occurs at the end of the design and construction period. It has the potential for greatly improving the efficient and effective use of Computerized Maintenance and Management System (CMMS) and Computer-Assisted Facility Management (CAFM) software applications by eliminating the tedious, time-consuming, and duplicative data re-collection effort that is often the first step in deploying these applications, a non-value-added task that is estimated to cost building owners a minimum of $50,000 per building.

The information structure at the heart of COBIE was initially developed by E. William (Bill) East, P.E., Ph.D., of the USACOE's Engineering Research and Development Center (ERDC).[7] While COBIE is designed to extract information directly from building information models, among the innovative features of Dr. East's design is that COBIE data may also be created and exchanged using simple spreadsheets, which allows anyone to use COBIE to compile structured information for any facility, regardless of the size, complexity, technological sophistication, and available capital of the business enterprise.

What should not be overlooked is that COBIE is not technology. It is a standard for structured information exchange that enables us to use technology more effectively. It is an excellent example of the type of vital innovation needed in the building industry that can lead to dramatic improvements in the efficiency, productivity, and quality of services delivered throughout the building life cycle. For more information about COBIE, visit the Whole Building Design Guide at www.wbdg.org.

 SPECIFIERS PROPERTY INFORMATION EXCHANGE (SPIE)

The COBIE project revealed a significant gap in information exchange related to information provided by building product manufacturers to design professionals. These information authors typically reduce editable alphanumeric data files to graphical images in Adobe Acrobat Portable Document Format (PDF),[8] making it completely inaccessible for electronic information exchange. This instance of electronic information degradation—with which design professionals are all too familiar—occurs well before the information can be used in even the earliest business processes in the building life cycle. Very little of it is rekeyed during design or construction, so most of the rich information generated by these authors is typically lost forever.

As a result, ERDC, together with the Specifications Consultants in Independent Practice (SCIP) and the Construction Specifications Institute (CSI), launched a new project sanctioned by the buildingSMART alliance to develop a specification to identify the full set of properties needed to specify materials, products, and equipment to a common, initial level of detail. SCIP has identified a total of 10,291 individual property sets that could be applied to products across the entire set of specification sections.[9] Figure 7.7 shows a software tool called a "Propertylizer" that is used to assign properties to specific materials, products, or equipment systematically. Adoption of this structured data format by building product and equipment manufacturers would relieve a tremendous bottleneck to information exchange that could greatly improve the efficiency of the building design process by enabling building product information to be incorporated directly into building information models. Benefits would continue to accrue throughout the construction, commissioning, and operations and maintenance phases of the building life cycle, where the structured format of COBIE is ready and waiting to receive such information.

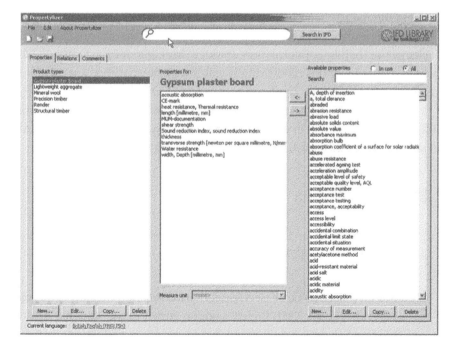

FIGURE 7.7
Propertylizer Tool for International Framework Dictionary Library. (Source: IFD Library Partners. Used by permission.)

COORDINATION VIEW INFORMATION EXCHANGE (CVIE)

The information management challenges of the building industry may seem overwhelming at times, but resolving these challenges is well within the realm of possibility, even probability. This chapter concludes with a rare building industry information exchange success story that illustrates that progress is possible.

It is widely accepted by design and construction professionals that one of the primary benefits of building information modeling is the ability to resolve physical interference (clash) problems virtually, thus eliminating the need to address these problems at far greater risk and expense during construction. This is what might be called the "coordination view" of the building information model, which became one of the first IFC Model View Definitions (MVDs) and has become the most common type of IFC data exchange.[10] As a result, "clash detection" has entered the building industry lexicon and is now considered an essential step of the building information modeling process during the building design phase. We now take clash detection for granted. Ironically, it involves only a small subset of data that most proprietary software applications can generate and that the IFC data model is capable of supporting, namely, building geometry. Additionally, the model view definition explicitly states that the round-trip transfer of BIM data is not part of the Coordination View business process. Yet this one-way, geometry-only information exchange is transforming the industry and having a significant positive impact on construction cost and schedules, which indicates just how great the potential for richer exchange of building information through building information modeling might be. Providing this report simply states to the owner that you certify that you have resolved all the conflicts in the model.

Many other projects are underway to remove the barriers to electronic information exchange. For a current, up-to-date list, visit the buildingSMART alliance Web site at www.buildingsmartalliance.org.

ENDNOTES

1. AEC3, "Information Delivery Manual – Norway," AEC3, http://www.aec3.com/5/5_009_IDM.htm.
2. Ibid.
3. Co-author Michael Tardif served as the Information Technology Project Manager for the agcXML Project under contract to the National Institute of Building Sciences.

4. International Code Council, "Smartcodes: ICC's Automated Code Checking Solution," (2008).

5. buildingSMART alliance, "Construction Operations Building Information Exchange," National Institute of Building Sciences, http://www.buildingsmartalliance.org/projects/projdetail.php?p=8.

6. E. William East, "Construction Operations Building Information Exchange (COBIE)," National Institute of Building Sciences, http://www.wbdg.org/resources/cobie.php.

7. _____,"Construction Operations Building Information Exchange (COBIE) Requirements Definition and Pilot Implementation Standard," no. ERDC/CERL TR-07-30 (2007), http://www.wbdg.org/pdfs/erdc_cerl_tr0730.pdf.

8. Ibid.

9. _____,"Specifiers' Properties Information Exchange (SPIE)," National Institute of Building Sciences, http://www.buildingsmartalliance.org/projects/projdetail.php?p=36.

10. _____, "Coordination View Information Exchange (CVIE)," National Institute of Building Sciences, http://www.buildingsmartalliance.org/projects/projdetail.php?p=37.

CHAPTER 8

The Way Forward

It doesn't work to leap a twenty-foot chasm in two
ten-foot jumps.

—American proverb

Many building industry professionals, accustomed to the fragmented nature of the building industry, remain unaware of the worldwide efforts being made to address the problems of workflow and building information exchange, only a few of which are described in the preceding chapter. The standards-development efforts with which most professionals are familiar— codes and standards—seem remote, highly formalized, and accessible only to a small subset of the industry. This has led to a business culture of limited involvement in collaborative, industry-wide efforts to advance the art and science of the industry, not because people are unwilling to participate, but rather because they don't quite know how, or simply don't realize that their participation is needed.

In the hundreds of speaking engagements in which we have presented these issues to professional audiences, we have observed a very consistent three-part reaction. The first part is that most professionals readily grasp the dimensions of the challenges facing the industry and the urgent need to address them. The second is a reflexive expectation that "someone" needs to solve these problems. The third is a general lack of awareness of the efforts under-way and the progress being made, or the time it takes to build a collaborative solution, which often results in entirely new, duplicative efforts.

We hope that this book has increased your awareness of the many activi-ties that are taking place to move the industry forward. We would also like to make an explicit pitch to contribute to those efforts. The "someone" who needs to work toward solving industry challenges is everyone—including you.

The degree of collaboration and knowledge sharing among building industry organizations worldwide has never been higher and is continually improving. If you wish to be part of the solution, you can do so through your own industry or professional organization, or you can join industry-wide organizations such as FIATECH or the buildingSMART alliance.

Regardless of the specific goals of any individual standards-development project, the overall goal of our efforts is to inspire as much confidence in build-ing information as we have in other types of digital information that we now handle on a regular basis, such as online banking, travel reservations, and shopping. Public confidence in e-Commerce transactions is achieved partly through the familiarity that comes with experience, but also through visibility, transparency, and accessibility of information. The interface for online bank-ing, credit card account access, and shopping is remarkably uniform from one business enterprise to another. Online banking is made transparent by allowing us to view our data in any number of ways: as a monthly statement, as a chro-nology of recent activity, or as individual transactions. For many banks, even facsimiles of paper transactions—cancelled checks—are readily accessible. We are provided with information about when pending transactions are expected to be executed—when a deposit will be available for withdrawal or when an online order will be shipped—and confirmation once they have been executed. It is even possible, in many cases, to monitor the status of many eCommerce transactions or activities in real time.

Just imagine if we had the same confidence in available data about the built world that is now beginning to occur in the geospatial world. Many communities are now able to retrieve information securely about a sewer line—its location and configuration in plan and its invert elevation—and can be completely confident that the information accurately depicts real-world conditions. Criti-cal to data integrity is that the information database interface includes a feedback loop, so that anyone who discovers errors in the database has a means for easily providing corrected information in the course of doing his or her job, which can then be validated by a responsible authority. We are, unfortunately, quite far from having that level of confidence, visibility, transparency, and accessibility of building information. This is the basis for much of the waste and inefficiency in our industry. Since we do not have a commonly accepted means of compiling and storing building information for future use, we are condemned endlessly to collecting the same information over and over again throughout the life of a building.

Poor performance is generally accepted as the norm for the building industry. Clients expect that activities will take longer and cost more than predicted, and build that negative expectation into their business plans. The workflows now in place have been used for decades and have simply become "the way we do business." While process improvements have been made and are being made in certain parts of the industry, the emergence of organizations such as the buildingSMART alliance, that view the industry holistically, is a very recent phenomenon.

WORKFLOW: FROM SEQUENTIAL TO PARALLEL PROCESSING

The fundamental problem of building industry workflow is that it is sequential, which means that gains in efficiency and productivity in one part of the sequence may be completely undermined by bottlenecks that occur before or after, and problems that remain unaddressed in one part of the sequence must be taken up in a subsequent part. Technology such as building information modeling and service delivery models such as Integrated Project Delivery have something very important in common: both foster business processes that can be executed in parallel rather than in sequence, thus opening the door to optimizing the entire business process instead of individual, sequential tasks. Workflow is redirected toward the final outcome. Individual tasks are evaluated on the basis of the value they add to the final product or service. Information is regarded less as a deliverable and more as a resource to be leveraged. The larger the data set, the more opportunities there are for scenario planning and iterative analysis—the "what if" game that can yield the tremendous insights that are generally unavailable to us now.

Data analyzed becomes information from which conclusions can be drawn to gain knowledge, which leads to the insight we call wisdom. It is a continuous and cumulative learning process that leads to progress. In the current state of the industry, we know desperately little about the connection between our actions and their consequences. Do buildings really perform as we have designed them to perform? Does a LEED Platinum building truly achieve LEED Platinum performance? Today, no one really knows. And because we don't know, we can't improve our methods, our standards, our technology, or our processes for creating the manmade world. The building industry as a whole must transform itself into a "learning organization," a model of the revolutionary management philosophy that Peter Senge, founder of the Center for Organizational Learning at MIT's Sloan School of Management, outlined in his groundbreaking book, *The Fifth Discipline.*

A learning organization undertakes a constant effort to evaluate and reevaluate its workflow and associated business processes and constantly improve and refine them. Each time one business process is improved, it will likely provide opportunities to improve other, associated business processes. A key aspect of success is understanding what information the disciplines that follow you need and how they need it. An architect or structural engineer may have no difficulty conceiving of a precast concrete façade that is ten stories tall. Both know that such a design is within the realm of possibility. But the contractor cannot erect a single, ten-story precast concrete facade. He or she needs to know how the unitary, finished result can best be assembled from a set of components. Yet, despite years of experience and even by intention, design professionals give little thought to the "methods and means of construction." Contractors, for their part, can be equally oblivious to their clients' operational needs, installing equipment without the needed clearances to service it properly and easily.

The most significant workflow issues are related to the use of information in a model for multiple purposes. Two parties working on different facets of the same building may have legitimate reasons to make conflicting adjustments. Siting decisions may be made by a geotechnical engineer for hydraulic reasons, while an architect or systems engineer may propose a different siting configuration for energy consumption reasons. These are dilemmas that can be resolved only by a joint exercise of professional judgment. No system of information management or analysis can automate these sorts of "trade-off" decisions. Cost is a constant trade-off factor, with the decision often predicated on qualitative factors. The best that can be hoped for from technology and streamlined business processes is that they will foster opportunities for these decisions to be made.

Many other business processes in the building industry, however, are ripe for full automation, if only the needed information can be properly structured and made available in the appropriate format. We accept, for example, that air travel, automobile rental, and hotel reservations can be made without human intervention. For the airline passenger, the name of the pilot and the tail number of the plane are irrelevant, but the airline needs to know and track those things to assure the passenger of a seat. If one aircraft must be replaced with another of a different configuration, the airline must know whether the pilot is certified to fly the replacement aircraft and must be able to notify the passengers of any changes in seating availability. Similarly comprehensive information databases and information model views are needed in the building industry. The online reservation system took many years to achieve its current optimum level of performance, but it very rapidly wiped out what consumers once regarded as a vital service but quickly came to regard as a non-value-added task: the trip or the phone call to a travel agent to make reservations and pick up tickets. The building industry is full of such non-value-added tasks and services that we now consider vital. If the airline industry, with all of its problems, can wring efficiency out of such a cumbersome and deeply entrenched business process, it should give us hope that we can do the same.

When we compare the building industry to other industries, the phenomenon of our broken business processes is really quite unacceptable, if not downright embarrassing. The CD-ROM industry got together very early in its development phase and agreed to a data encryption format called the High Sierra Standard, named for the hotel where it was adopted. A similar agreement was made on infrastructure for the cell phone business. Anyone using any cell phone can talk to anyone using any other cell phone. The standards adopted by these industries left plenty of room for individual companies to distinguish themselves in the marketplace with proprietary features.

Figure 8.1 identifies the various layers of information exchange in the building industry. Some of the foundational tools we rely upon, such as the Internet, are ubiquitous. Other technologies need to be developed to meet the specific needs of our industry. The attributes of the metadata layer is an element of technology that we can and must agree to. We need to ensure that we are in agreement as to the attributes of the objects we exchange, which will enable us to use the metadata as a "security key" to determine who should be authorized to access the information beyond the metadata level.

The next layer, spatial communication, is an information layer only now emerging as a focus of concern in the building industry. An artificial boundary has developed between information that is inside versus outside a building. The split occurred because of the inability of computers and software applications to handle both environments at the same time. These limitations are quickly vanishing, and players on both sides of the divide are eyeing business opportunities in the data set on the other side. We have an opportunity to address the issue in a spirit of collaboration and common purpose—or we can continue the fragmented "go it alone" approach that has been the hallmark of technology development in the building industry for thirty years. Most of our current development work described in this book is taking place at the infrastructure layer. This is where most of the workflow development is being accomplished.

FIGURE 8.1
Information Management
Layers in the Building
Industry.

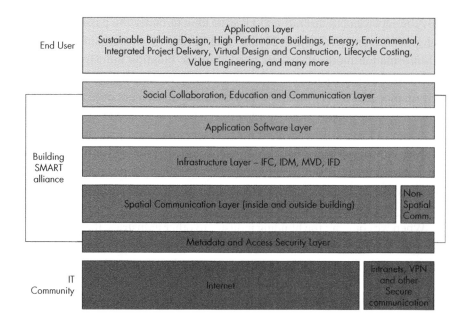

Information standards become less critical at the final three layers, while interoperability—the foundation for which should be laid at the infrastructure layer—becomes exponentially more important. Beyond technology, the building industry needs to establish a repository for best business practices, which could be modeled on the Information Technology Infrastructure Library (ITIL) created by the IT industry.

BUSINESS AND CONTRACTUAL RELATIONSHIPS

Changes in business processes will inevitably result in significant changes of business and contractual relationships. We must take advantage—and not fear—the opportunities that this element of change provides. To restrain a new way of doing business with outdated forms of business relationships would be not only foolish but costly, and would keep us from realizing the full potential of BIM.

One early effect on business relationships in the design and construction sector of the industry can be attributed to uneven implementation and acceptance of change. Though architects were the first building professionals to explore BIM technology, the pace of adoption by architects has been painfully slow, in part due to market conditions that have been extremely favorable to architects for nearly a decade, and in part out of concern over liability and compensation issues. Contractors, on the other hand, though relatively late to the game, are embracing BIM enthusiastically and in droves. Interestingly, not a single contractor, in our experience, has expressed a desire to encroach on the architect's realm of professional expertise—design. On the contrary, most explicitly disclaim any interest in assuming greater design responsibilities, fully cognizant of the fact that design is outside their realm of expertise. Rather, they are adopting BIM because it allows them to do what they do better—build buildings. But unless architects accelerate their adoption and implementation of BIM, the "rules of engagement" between architects and contractors are likely to change in a way that will be unfavorable to architects. BIM opens the door to ample, even remarkable business opportunities for those architects with the insight to perceive and exploit them.

Changes in business relationships can be anticipated to occur at every juncture, whether the parties are architects, engineers, contractors, subcontractors, fabricators, or building product and equipment suppliers. Access to real time information will become the norm sooner or later, and decision makers will need to be interacting constantly with the information and making decisions in an environment of collaboration, cooperation, and constantly changing factors.

EVOLVING ROLES AND RESPONSIBILITIES

Oddly, or perhaps reassuringly, roles and responsibilities of individual players in the building industry are not likely to change fundamentally as a result of BIM. Each player retains a distinct realm of expertise vital to the industry as a whole. It is the nature of the relationships that will change, becoming more collaborative and less adversarial or sequential. As more things occur in parallel, better communication will need to occur and will occur. A sports analogy is useful here. In any team sport, if the team is communicating then the team is much more likely to win. If, on the other hand, a quarterback throws a pass and the receiver is not running the correct, agreed-upon pattern, then the pass (or transaction) will not be complete. After a few failed passes (transactions), the game (business process) is lost. In the building industry, as in team sports, each player has a job to do. Assembling a team without an understanding of the roles and responsibilities of each player simply will not yield positive results.

Examining our business processes to determine how we can improve workflow is not much different than a team coach developing and implementing new plays. The goal is not to complicate our work with endless introspection and analysis. Instead, we want to examine and implement options that allow us to achieve peak performance and effectively move the ball toward the goal: building the building more efficiently and truly meeting the needs of our clients.

Bibliography

AEC3. "Information Delivery Manual: Norway." Retrieved from http://www
.aec3.com/5/5_009_IDM.htm.

Alliance of Automobile Manufacturers. "Recycling Vehicles." Alliance of
Automobile Manufacturers. Retrieved from http://www.autoalliance.org/
environment/recycling.php.

American Heritage Dictionary of the English Language. 4th ed. Houghton Mifflin Company, 2004.

Barton, S., Ed. *Capital Projects Technology Roadmap.* Austin, Texas: FIATECH,
2004.

buildingSMART alliance. "Construction Operations Building Information
Exchange." National Institute of Building Sciences. Retrieved from http://
www.buildingsmartalliance.org/projects/projdetail.php?p=8.

Bureau of Labor Statistics. "Industry at a Glance: Naics 23: Construction."
U.S. Department of Labor. Retrieved from http://www.bls.gov/iag/con-
struction.htm.

——— "Occupational Outlook Handbook, 2008–09 Edition." U.S. Department of Labor. Retrieved from http://www.bls.gov/oco/.

Dictionary.com. "Wordnet 3.0." Princeton University. Retrieved from http://
dictionary.reference.com/browse/information.

Diekmann, James E., Mark Krewedl, Joshua Balonick, Travis Stewart, and
Spencer Wonis. "Application of Lean Manufacturing Principles to Construction." Austin, Tex.: Construction Industry Institute, 2004.

East, E. William. "Construction Operations Building Information Exchange
(COBIE)." National Institute of Building Sciences. Retrieved from http://
www.wbdg.org/resources/cobie.php.

——— "Construction Operations Building Information Exchange (COBIE)
Requirements Definition and Pilot Implementation Standard," no. ERDC/
CERL TR-07–30 (2007). Retrieved from http://www.wbdg.org/pdfs/
erdc_cerl_tr0730.pdf.

——— "Coordination View Information Exchange (CVIE)." National Institute
of Building Sciences. Retrieved from http://www.buildingsmartalliance
.org/projects/projdetail.php?p=37.

———— "Specifiers' Properties Information Exchange (SPIE)." National Institute of Building Sciences. Retrieved from http://www.buildingsmartalliance .org/projects/projdetail.php?p=36.

Eastman, Charles, Seok-Joon You, Frank Wang, Donghoon Yang, and Jae-Min Lee. "CIS/2 Exchange Capabilities: Translator Interchangeability Matrix." Georgia Institute of Technology. Retrieved from http://www.coa.gatech .edu/-aisc/index.php?cat1=1.

FIATECH. (2007). "BIM for Precast Concrete." Retrieved August 7, 2008, from http://www.fiatech.org/projects/autodesign/bpc.htm#_projpartact.

FMI and Construction Management Association of America. "Eighth Annual Survey of Owners," 2007 Retrieved July 29, 2008, from http://www.fmire-sources.com/pdfs/07SOA.pdf.

Federal Facilities Council. "Federal Facilities Beyond the 1990s: Ensuring Quality in an Era of Limited Resources." Washington, DC Federal Facilities Council, 1997.

Fuglie, Keith O., James M. MacDonald, and Eldon Ball. "Productivity Growth in U.S. Agriculture." edited by United States Department of Agriculture. Washington, D.C.: Economic Research Service, 2007.

Fuhrman, Andy. "Oscre Update: The OSCRE Standard for Property E-Commerce." In *American Institute of Architects Building Connections 3.* Washington, DC: American Institute of Architects, 2006.

Haskell, Preston H. "Construction Industry Productivity: Its History and Future Direction." In *The Haskell Company White Papers.* Jacksonville, Florida: Haskell Company, 2004.

International Code Council. "SMARTcodes: ICC's Automated Code Checking Solution," 2008.

Khemlani, Lachmi. "The CIS/2 Format: Another AEC Interoperability Standard." AECbytes. Retrieved from http://www.aecbytes.com/build-ingthefuture/2005/CIS2format.html.

Knowledge Based Systems. "IDEF: Integrated Definition Models," 2006. Retrieved August 7, 2008, from http://www.idef.com.

Lubowski, Robert N., Marlow Vesterby, Shawn Bucholtz, Alba Baez, and Michael J. Roberts. "Major Uses of Land in the United States, 2002." edited by United States Department of Agriculture. Washington, D.C.: Economic Research Service, 2006.

Matos, Grecia, and Lorie Wagner. "Consumption of Materials in the United States, 1900–1995." Annual Review of Energy and the Environment 1998, no. 23 (1998): 107–22.

Mazria, Edward, and Kristina Kershner. *The 2030 Blueprint: Solving Climate Change Saves Billions.* Santa Fe, N.M.: Architecture 2030, 2008.

National BIM Standard Project Committee. "National Bim Standard Capability Maturity Model Workbook." National Institute of Building Sciences. Retrieved from http://www.facilityinformationcouncil.org/bim/I-CMM.

———— National Building Information Modeling Standard, Version 1, Part 1: Overview, Principles, and Methodologies, 2007. Washington, DC, National Institute of Building Sciences. Retrieved from http://www.facilityinformationcouncil.org/bim/publications.php, ed, 2007.

National Institute of Standards and Technology. "CIS/2 and IFC: Product Data Standards for Structural Steel." National Institute of Standards and Technology. Retrieved from http://cic.nist.gov/vrml/cis2.html.

North Carolina Department of Cultural Resources. "Best Practices for File-Naming," 2008. Retrieved from http://www.records.ncdcr.gov/e_records/filenaming_20080508_final.pdf.

Object Management Group. "Business Process Modeling Notation (BPMN) Information," 2007. Retrieved August 7, 2008, from http://www.bpmn.org.

"Obsolete Card Catalog Files at Sterling Memorial Library, Yale University." Wikimedia Commons. Retrieved from http://commons.wikimedia.org/wiki/Image:Yale_card_catalog.jpg.

OCCS Development Committee Secretariat. "OmniClass: A Strategy for Classifying the Built Environment." Construction Specifications Institute. Retrieved from http://www.omniclass.org/index.asp.

Office of Integrated Analysis and Forecasting. "Annual Energy Outlook 2007." Washington, D.C.: Energy Information Administration, U.S. Department of Energy, 2007.

Population Division of the Department of Economic and Social Affairs. "World Population Prospects: The 2004 Revision." United Nations Secretariat, 2005.

Roodman, David Malin, and Nicholas Lenssen. Worldwatch Paper #124: A Building Revolution: How Ecology and Health Concerns Are Transforming Construction." Worldwatch Institute, 1995.

Teicholz, Paul. "Labor Productivity Declines in the Construction Industry: Causes and Remedies." AECBytes, April 14, 2004. Retrieved from http://www.aecbytes.com/viewpoint/2004/issue_4.html.

United States Department of Agriculture. "Agricultural Productivity in the United States." Economic Research Service. Retrieved from http://www.ers.usda.gov/data/agproductivity/.

U.S. Green Building Council. "Green Building Facts." Washington, D.C., 2008.

———— "LEED Frequently Asked Questions," 2008. Retrieved August 11, 2008, from http://www.usgbc.org/DisplayPage.aspx?CMSPageID=1819#LEED.

Wendell Cox Consultancy. "U.S. Population from 1900." Retrieved from http://www.demographia.com/db-uspop1900.htm.

Wikimedia Commons. "Swiss Army Knife Wenger Opened." Retrieved from http://commons.wikimedia.org/wiki/Image:Swiss_Army_Knife_Wenger_Opened_20050627.jpg.

Wikipedia. "Barcode," 2008. Retrieved August 5, 2008, from http://en.wikipedia.org/wiki/Barcode.

———Wikipedia. "Capability Maturity Model." Retrieved from http://en.wikipedia.org/wiki/Capability_Maturity_Model.

——— "Database Normalization." Retrieved from http://en.wikipedia.org/wiki/Database_normalization.

——— "ISO 10303." Retrieved from http://en.wikipedia.org/wiki/ISO_10303.

——— "Service-Oriented Architecture." Retrieved from http://en.wikipedia.org/wiki/Service-oriented_architecture.

Index

A

Adobe Acrobat, 166

AGC. *See* Associated General Contractors of America

agcXML, 69, 151, 161, 162, 168n

AIA. See American Institute of Architects

Airbus, 4

AISC. *See* American Institute of Steel Construction

American Institute of Architects (AIA), 93

    AIA Contract Documents, 137

    AIA C106-2007, Digital Data License Agreement, 137

    AIA E201-2007, Digital Data Protocol, 137

    AIA E202-2008, BIM Protocol Exhibit, 137

    AIA Technology in Architecture Practice (TAP), 46

American Institute of Steel Construction (AISC), 150, 151, 152n

API. *See* Application Programming Interface

Application Programming Interface (API), 151

ArchiCAD. *See* Graphisoft ArchiCAD

Arkansas Precast, 82

Associated General Contractors of America (AGC), 69, 137, 151, 161

architecture, 9, 82, 89, 108, 117

Architecture 2030, 9, 25n

AutoCAD. See Autodesk AutoCAD

Autodesk

AutoCAD Architecture, 38t

Revit, 18, 38t

Navisworks, 18

B

BIM Protocol Exhibit. See American Institute of Architects

Beck Technology

    dProfiler, 18

Bentley Systems

    Bentley Architecture, 18, 38t

    Bentley Building Electrical Systems, 38t

    Bentley Building Mechanical Systems, 38t

    Bentley Structural, 38t

BLS. *See* Bureau of Labor Statistics

Boeing, 4

buildingSMART

    buildingSMART alliance, 68, 69, 79, 82, 117, 151, 161, 163, 167, 168, 169n, 172, 173

    buildingSMART International, 39, 68, 79, 151, 154, 164

Bureau of Labor Statistics, 5, 6, 6f, 25n, 101, 110n

C

CAD. *See* Computer-aided Design

CAFM. *See* Computer-aided Facility Management

Center for Integrated Facility Engineering (CIFE), 6f

CIFE. *See* Center for Integrated Facility Engineering
CII. *See* Construction Industry Institute
CIMSteel Integration Standards, 150, 151, 152
CIS/2. *See* CIMSteel Integration Standards
clash detection, 18, 37, 103, 118, 142, 168
collaboration, 35, 43, 57, 65, 106, 109, 110, 111, 112, 141, 142, 172, 175, 176, 177
Computer-aided Design (CAD), 14, 28, 33, 40, 82, 101, 106, 107, 142, 145
Computer-aided Facility Management (CAFM), 166
concrete, 11, 38, 80, 82, 83, 84, 85, 86, 88, 92, 154, 160, 174
ConsensusDOCS, 93, 137
Construction Industry Institute (CII), 10, 69
Construction Specifications Institute (CSI), 167
CSI. *See* Construction Specifications Institute

D
design-build, 89
Digital Data License Agreement. *See* American Institute of Architects
Digital Data Protocol. *See* American Institute of Architects
dProfiler. *See* Beck Technology

E
E201-2007. *See* American Institute of Architects
E202-2008. *See* American Institute of Architects
Ecotect, 18
estimating, 18, 62, 63, 86, 138, 139, 148, 162
Excel. *See* Microsoft Excel

F
FIATECH, 69, 70, 71, 72, 82, 88, 172

G
General Services Administration. *See* U.S. General Services Administration
Geographical Information Systems (GIS), 45t, 51, 53t
Georgia Institute of Technology, 82, 150, 152
Georgia Tech. *See* Georgia Institute of Technology
Gehry, Frank, 107, 108
Gehry Technologies, 108
GIS. *See* Geographical Information Systems
Global Positioning System (GPS), 15, 53t, 145t
GPS. *See* Global Positioning System
Graphisoft, 113f
Graphisoft ArchiCAD, 18, 38t
green building, 9, 25, 67, 88
GSA. *See* U.S. General Services Administration

H
HKS, Inc., 82
High Concrete Structures, Inc., 82
human resources, 98, 100, 101, 105
HVAC (heating, ventilation and air-conditioning), 38t, 92, 122

I
IAI. *See* International Alliance for Interoperability
ICC. *See* International Code Council
IDM. *See* Information Delivery Manual
IES VE-Ware, 18
IFC. *See* Industry Foundation Classes
IFD. *See* International Framework Dictionary
Industry Foundation Classes (IFC), 38, 39, 45t, 54, 54t, 68, 69, 85f, 149, 151, 152n, 154, 155, 164, 168, 176f
Information Delivery Manual (IDM), 117, 118, 154, 155, 157, 168n, 176f
Intelisum, 19

intelligent objects, 148t

International Alliance for Interoperability (IAI), 39, 68

International Code Council (ICC), 117, 162, 169

International Framework Dictionary (IFD), 167f

interference checking. *See* clash detection

interoperability, 8, 33, 39, 45, 54, 65, 66, 68, 69, 144, 146, 147, 149, 152, 159, 176

L

Labor productivity, 4, 5, 6, 7, 13, 25n, 77, 101, 102n, 103, 107

laser scanning, 19

Leadership in Energy and Environmental Design (LEED), 18, 67, 88n, 113f, 118, 148t, 157, 173

lean construction, 25n

LEED. *See* Leadership in Energy and Environmental Design

M

mechanical, electrical, and plumbing (MEP), 38t

MEP. *See* mechanical, electrical, and plumbing

Microsoft

    Excel, 62, 84

    Visio, 82

Model View Definition (MVD), 117, 118, 119, 168, 176f

MVD. *See* Model View Definition

N

NASA. *See* National Aeronautics and Space Administration

National Aeronautics and Space Administration (NASA), 22, 163, 164

National BIM Standard (NBIMS), 44, 46, 69, 70, 72, 139

National CAD Standard (NCS), 45t, 52, 97

National Institute of Building Sciences (NIBS), 55n, 68, 69, 88n, 161, 163, 168n, 169n

National Institute of Standards and Technology (NIST), 152n

Navisworks. *See* Autodesk, Navisworks

NBIMS. *See* National BIM Standard

NCS. *See* National CAD Standard

Newforma Project Center, 40, 133

NIBS. *See* National Institute of Building Sciences

NIST. *See* National Institute of Standards and Technology

O

Omniclass, 146, 151n

Onuma, Kimon, 62, 103, 127n

Onuma Planning System (OPS), 19, 62, 116, 124f, 125, 126f,

OPS. *See* Onuma Planning System

P

parametric, 51

precast concrete, 38t, 80f, 82, 83f, 84f, 85f, 86, 88n, 154, 174

prefabrication, 63

procurement, 69, 70, 71f, 100f, 113f, 145t, 147t, 150

Q

QA. *See* quality assurance

Quality assurance, 97

Quantapoint, 19

quantity takeoff, 92

R

radio-frequency identification (RFID), 66, 74

Revit. *See* Autodesk Revit

RFID. *See* radio-frequency identification

S

safety, 14, 79, 114, 144, 148t

service-oriented architecture (SOA), 45t, 50, 51t, 144

shop drawings, 63, 83

SMARTcodes, 117, 162, 163, 169n

SOA. *See* service-oriented architecture
Solibri, 18, 38t
SPECSINTACT, 164
steel, 3, 12, 38, 80f, 150, 152n
supply chain management, 3, 4, 5, 12, 21, 65,
    66, 67, 143, 144, 146

T
takeoff. *See* quantity takeoff
TAP. *See* American Institute of Architects
Technology in Architectural Practice (TAP). *See*
    American Institute of Architects
Tekla Structures, 38t

U
user interface, 31, 54t, 125, 146, 163, 172
USGBC. *See* U.S. Green Building Council
U.S. General Services Administration
    (GSA), 154
U.S. Green Building Council (USGBC), 25n,
    67, 88n

V
value engineering, 176f
virtual design and construction, 176f